Speed Maths
for Kids

Helping Children Achieve
Their Full Potential

Bill Handley

WILEY

First published in 2005 by Wrightbooks
an imprint of John Wiley & Sons Australia, Ltd
42 McDougall St, Milton Qld 4064

Office also in Melbourne

Reprinted with updates 2022

Typeset in 11.5/15 pt Plantin Std

© Bill Handley 2005

The moral rights of the author have been asserted

ISBN: 978-0-731-40227-4

A catalogue record for this book is available from the National Library of Australia

Cover design by Wiley
Cover image: © Milko/Getty Images

Disclaimer

Contents

Preface

I could have called this book *Fun With Speed Mathematics*. It contains some of the same material as my other books and teaching materials. It also includes additional methods and applications based on the strategies taught in *Speed Mathematics* that, I hope, give more insight into the mathematical principles and encourage creative thought. I have written this book for younger people, but I suspect that people of any age will enjoy it.

In this revision of *Speed Maths for Kids*, I have included extra material on calculating percentages, an easy method of long division you can perform mentally, calculating from left to right, and an easy method for multiplying when other methods don't seem to work. I trust that these extra sections will make this an exciting and enjoyable read.

A common response I hear from people who have read my books or attended a class is, 'Why wasn't I taught this at school?' People feel that, with these methods, mathematics would have been so much easier, and they could have achieved better results than they did, or they feel they would have enjoyed mathematics a lot more. I would like to think this book will help on both counts.

I have definitely not intended *Speed Maths for Kids* to be a serious textbook but rather a book to be played with and enjoyed. I have written this book in the same way that I speak to young students. Some of the language and terms I have used are definitely non-mathematical. I have tried to write the book primarily so readers will understand. A lot of my teaching in the classroom has just been explaining out loud what goes on in my head when I am working with numbers or solving a problem.

I have received mail from people around the world who have enjoyed my book and found it helpful. I have also been gratified to learn that many schools are using my methods. I receive emails every day from students and teachers who are becoming excited about mathematics. I have included sections throughout this book to provide extra guidance for parents and teachers. I have also produced a handbook for teachers with instructions for teaching these methods in the classroom and with handout sheets for photocopying. Please email me or visit my website for details.

My sincere wish is that this book will inspire readers to enjoy mathematics and show them that they are capable of greatness.

Bill Handley
Melbourne, June 2022
bhandley@speedmathematics.com
www.speedmathematics.com

Introduction

I have heard many people say they hate mathematics. I don't believe them. They *think* they hate mathematics. It's not really maths they hate; they hate failure. If you continually fail at mathematics, you will hate it. No-one likes to fail.

But if you succeed and perform like a genius you will love mathematics. Often, when I visit a school, students will ask their teacher, can we do maths for the rest of the day? The teacher can't believe it. These are kids who have always said they hate maths.

If you are good at maths, people think you are smart. People will treat you like you are a genius. Your teachers and your friends will treat you differently. You will even think differently about yourself. And there is good reason for it — if you are doing things that only smart people can do, what does that make you? Smart!

I have had parents and teachers tell me something very interesting. Some parents have told me their child just won't try when it comes to mathematics. Sometimes they tell me their child is lazy. Then the child has attended one of my classes or read my books. The child not only does much better in maths, but also works much harder.

Why is this? It is simply because the child sees results for his or her efforts.

Often parents and teachers will tell the child, 'Just try. You are not trying.' Or they tell the child to try harder. This just causes frustration. The child would like to try harder but doesn't know how. Usually children just don't know where to start. Sometimes they will screw up their face and hit the side of their head with their fist to show they are trying, but that is all they are doing. The only thing they accomplish is a headache. Both child and parent become frustrated and angry.

I am going to teach you, with this book, not only what to do but how to do it. *You can be a mathematical genius.* You have the ability to perform lightning calculations in your head that will astonish your friends, your family and your teachers. This book is going to teach you how to perform like a genius — to do things your teacher, or even your principal, can't do. How would you like to be able to multiply big numbers or do long division in your head? While the other kids are writing the problems down in their books, you are already calling out the answer.

The kids (and adults) who are geniuses at mathematics don't have better brains than you — they have better methods. This book is going to teach you those methods. I haven't written this book like a schoolbook or textbook. This is a book to play with. You are going to learn easy ways of doing calculations, and then we are going to play and experiment with them. We will even show off to friends and family.

When I was in Year 9 I had a mathematics teacher who inspired me. He would tell us stories of Sherlock Holmes or of thriller movies to illustrate his points. He would often say, 'I am not supposed to be teaching you this,' or, 'You are not supposed to learn this for another year or two.' Often I couldn't wait to get home from school to try more examples for myself. He didn't teach mathematics like the other teachers. He told stories and taught us short cuts that would help us beat the other classes. He made maths exciting. He inspired my love of mathematics.

When I visit a school I sometimes ask students, 'Who do you think is the smartest kid in this school?' I tell them I don't want to know the person's name. I just want them to think about who the person is. Then I ask, 'Who thinks that the person you are thinking of has been told they are stupid?' No-one seems to think so.

Everyone has been told at one time that they are stupid — but that doesn't make it true. We all do stupid things. Even Einstein did stupid things, but he wasn't a stupid person. But people make the mistake of thinking that this means they are not smart. This is not true; highly intelligent people do stupid things and make stupid mistakes. I am going to prove to you as you read this book that you are very intelligent. I am going to show you how to become a mathematical genius.

How to read this book

Read each chapter and then play and experiment with what you learn before going to the next chapter. Do the exercises— don't leave them for later. The problems are not difficult. It is only by solving the exercises that you will see how easy the methods really are. Try to solve each problem in your head.

You can write down the answer in a notebook. Find yourself a notebook to write your answers and to use as a reference. This will save you writing in the book itself. That way you can repeat the exercises several times if necessary. I would also use the notebook to try your own problems.

Remember, the emphasis in this book is on playing with mathematics. Enjoy it. Show off what you learn. Use the methods as often as you can. Use the methods for checking answers every time you make a calculation. Make the methods part of the way you think and part of your life.

Now, go ahead and read the book and make mathematics your favourite subject.

Chapter 1
Multiplication: Getting started

How well do you know your multiplication tables? Do you know them up to the 15 or 20 times tables? Do you know how to solve problems like 14 × 16, or even 94 × 97, without a calculator? Using the speed mathematics method, you will be able to solve these types of problems in your head. I am going to show you a fun, fast and easy way to master your tables and basic mathematics in minutes. I'm not going to show you how to do your tables the usual way. The other kids can do that.

Using the speed mathematics method, it doesn't matter if you forget one of your tables. Why? Because if you don't know an answer, you can simply do a lightning calculation to get an instant solution. For example, after showing her the speed mathematics methods, I asked eight-year-old Trudy, 'What is 14 times 14?' Immediately she replied, '196.'

I asked, 'You knew that?'

She said, 'No, I worked it out while I was saying it.'

Would you like to be able to do this? It may take five or ten minutes' practice before you are fast enough to beat your friends even when they are using a calculator.

What is multiplication?

How would you add the following numbers?

$6 + 6 + 6 + 6 + 6 + 6 + 6 + 6 = ?$

You could keep adding sixes until you get the answer. This takes time and, because there are so many numbers to add, it is easy to make a mistake.

The easy method is to count how many sixes there are to add together, and then use multiplication tables to get the answer.

How many sixes are there? Count them.

There are eight.

You have to find out what eight sixes added together would make. People often memorise the answers or use a chart, but you are going to learn a very easy method to calculate the answer.

As a multiplication, the problem is written like this:

$8 \times 6 =$

This means there are eight sixes to be added. This is easier to write than $6 + 6 + 6 + 6 + 6 + 6 + 6 + 6 =$.

The solution to this problem is:

$8 \times 6 = 48$

The speed mathematics method

I am now going to show you the speed mathematics way of working this out. The first step is to draw circles under each of the numbers. The problem now looks like this:

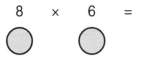

We now look at each number and ask, how many more do we need to make 10?

We start with the 8. If we have 8, how many more do we need to make 10?

The answer is 2. Eight plus 2 equals 10. We write 2 in the circle below the 8. Our equation now looks like this:

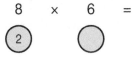

We now go to the 6. How many more to make 10? The answer is 4. We write 4 in the circle below the 6.

This is how the problem looks now:

We now take away crossways, or diagonally. We either take 2 from 6 or 4 from 8. It doesn't matter which way we subtract, the answer will be the same, so choose the calculation that looks easier. Two from 6 is 4, or 4 from 8 is 4. Either way the answer is 4. You only take away one time. Write 4 after the equals sign.

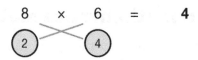

For the last part of the answer, you 'times' the numbers in the circles. What is 2 times 4? Two times 4 means two fours added together. Two fours are 8. Write the 8 as the last part of the answer. The answer is 48.

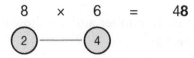

Easy, wasn't it? This is much easier than repeating your multiplication tables every day until you remember them. And this way, it doesn't matter if you forget the answer, because you can simply work it out again.

Do you want to try another one? Let's try 7 times 8. We write the problem and draw circles below the numbers like before:

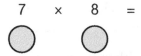

How many more do we need to make 10? With the first number, 7, we need 3, so we write 3 in the circle below the 7. Now go to the 8. How many more to make 10? The answer is 2, so we write 2 in the circle below the 8.

Our problem now looks like this:

Now take away crossways. Either take 3 from 8 or 2 from 7. Whichever way we do it, we get the same answer. Seven minus 2 is 5 or 8 minus 3 is 5. Five is our answer either way. Five is the first digit of the answer. You only do this calculation once so choose the way that looks easier.

The calculation now looks like this:

$$7 \quad \times \quad 8 \quad = \quad \mathbf{5}$$
$$\textcircled{3} \qquad \textcircled{2}$$

For the final digit of the answer we multiply the numbers in the circles: 3 times 2 (or 2 times 3) is 6. Write the 6 as the second digit of the answer.

Here is the finished calculation:

$$7 \quad \times \quad 8 \quad = \quad \mathbf{56}$$
$$\textcircled{3} \qquad \textcircled{2}$$

Seven eights are 56.

How would you solve this problem in your head? Take both numbers from 10 to get 3 and 2 in the circles. Take away crossways. Seven minus 2 is 5. We don't say five, we say, 'Fifty … '. Then multiply the numbers in the circles. Three times 2 is 6. We would say, 'Fifty … six.'

With a little practice you will be able to give an instant answer. And, after calculating 7 times 8 a dozen or so times, you will find you remember the answer, so you are learning your tables as you go.

Test yourself

Here are some problems to try by yourself. Do all of the problems, even if you know your tables well. This is the basic strategy we will use for almost all of our multiplication.

(a) 9 × 9 =

(e) 8 × 9 =

(b) 8 × 8 =

(f) 9 × 6 =

(c) 7 × 7 =

(g) 5 × 9 =

(d) 7 × 9 =

(h) 8 × 7 =

How did you go? The answers are:

| (a) 81 | (b) 64 | (c) 49 | (d) 63 |
| (e) 72 | (f) 54 | (g) 45 | (h) 56 |

Isn't this the easiest way to learn your tables?

Now, cover your answers and do them again in your head. Let's look at 9 × 9 as an example. To calculate 9 × 9, you have 1 below 10 each time. Nine minus 1 is 8. You would say, 'Eighty ...'. Then you multiply 1 times 1 to get the second half of the answer, 1. You would say, 'Eighty ... one.'

If you don't know your tables well it doesn't matter. You can calculate the answers until you do know them, and no-one will ever know.

Multiplying numbers just below 100

Does this method work for multiplying larger numbers? It certainly does. Let's try it for 96 × 97.

96 × 97 =

What do we take these numbers up to? How many more to make what? How many to make 100, so we write 4 below 96 and 3 below 97.

What do we do now? We take away crossways: 96 minus 3 or 97 minus 4 equals 93. Write that down as the first part of the answer. What do we do next? Multiply the numbers in the circles: 4 times 3 equals 12. Write this down for the last part of the answer. The full answer is 9,312.

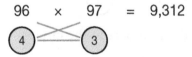

Which method do you think is easier, this method or the one you learnt in school? I definitely think this method; don't you agree?

Let's try another. Let's do 98 × 95.

98 × 95 =

First we draw the circles.

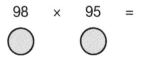

How many more do we need to make 100? With 98 we need 2 more and with 95 we need 5. Write 2 and 5 in the circles.

$$98 \quad \times \quad 95 \quad =$$

(2)   (5)

Now take away crossways. You can do either 98 minus 5 or 95 minus 2.

$98 - 5 = 93$

or

$95 - 2 = 93$

The first part of the answer is 93. We write 93 after the equals sign.

$$98 \quad \times \quad 95 \quad = \quad \textbf{93}$$

(2)   (5)

Now multiply the numbers in the circles.

$2 \times 5 = 10$

Write 10 after the 93 to get an answer of 9,310.

$$98 \quad \times \quad 95 \quad = \quad 9{,}3\textbf{10}$$

(2)   (5)

Easy. With a couple of minutes' practice you should be able to do these in your head. Let's try one now.

$96 \times 96 =$

In your head, draw circles below the numbers.

What goes in these imaginary circles? How many to make 100? Four and 4. Picture the equation inside your head. Mentally write 4 and 4 in the circles.

Now take away crossways. Either way you are taking 4 from 96. The result is 92. You would say, 'Nine thousand, two hundred ... '. This is the first part of the answer.

Now multiply the numbers in the circles: 4 times 4 equals 16. Now you can complete the answer: 9,216. You would say, 'Nine thousand, two hundred ... and sixteen.'

This will become very easy with practice.

Try it out on your friends. Offer to race them and let them use a calculator. Even if you aren't fast enough to beat them you will still earn a reputation for being a brain.

Beating the calculator

To beat your friends when they are using a calculator, you only have to start calling the answer before they finish pushing the buttons. For instance, if you were calculating 96 times 96, you would ask yourself how many to make 100, which is 4, and then take 4 from 96 to get 92. You can then start saying, 'Nine thousand, two hundred ...'. While you are saying the first part of the answer you can multiply 4 times 4 in your head, so you can continue without a pause, '... and sixteen.'

You have suddenly become a maths genius!

Test yourself

Here are some more problems for you to do by yourself.

(a)   96 × 96 =

(e)   98 × 94 =

(b)   97 × 95 =

(f)   97 × 94 =

(c)   95 × 95 =

(g)   98 × 92 =

(d)   98 × 95 =

(h)   97 × 93 =

The answers are:

| (a) 9,216 | (b) 9,215 | (c) 9,025 | (d) 9,310 |
| (e) 9,212 | (f) 9,118 | (g) 9,016 | (h) 9,021 |

Did you get them all right? If you made a mistake, go back and find where you went wrong and try again. Because the method is so different, it is not uncommon to make mistakes at first.

Are you impressed?

Now, do the last exercise again, but this time, do all of the calculations in your head. You will find it much easier than you imagine. You need to do at least three or four calculations in your head before it really becomes easy. So, try it a few times before you give up and say it is too difficult.

I showed this method to a boy in first grade and he went home and showed his dad what he could do. He multiplied 96 times 98 in his head. His dad had to get his calculator out to check if he was right!

Keep reading, and in the next chapters you will learn how to use the speed maths method to multiply any numbers.

Chapter 2
Using a reference number

In this chapter we are going to look at a small change to the method that will make it easy to multiply any numbers.

Reference numbers

Let's go back to 7 times 8:

$$7 \times 8 =$$

The 10 at the left of the problem is our *reference number*. It is the number we subtract the numbers we are multiplying from.

The reference number is written to the left of the problem. We then ask ourselves, is the number we are multiplying *above* or *below* the reference number? In this case, both numbers are below, so we put the circles below the numbers. How many below 10 are they? Three and 2.

We write 3 and 2 in the circles. Seven is 10 minus 3, so we put a minus sign in front of the 3. Eight is 10 minus 2, so we put a minus sign in front of the 2.

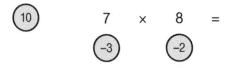

We now take away crossways: 7 minus 2 or 8 minus 3 is 5. We write 5 after the equals sign.

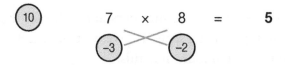

Now, here is the part that is different. We multiply the 5 by the reference number, 10. Five times 10 is 50, so write a 0 after the 5. (How do we multiply by 10? Simply put a 0 at the end of the number.) Fifty is our subtotal. Here is how our calculation looks now:

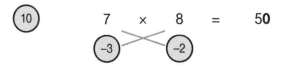

Now multiply the numbers in the circles. Three times 2 is 6. Add this to the subtotal of 50 for the final answer of 56.

The full working out looks like this:

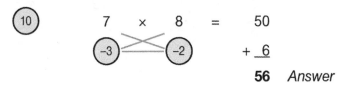

Why use a reference number?

Why not use the method we used in Chapter 1? Wasn't that easier? That method used 10 and 100 as reference numbers as well, we just didn't write them down.

Using a reference number allows us to calculate problems such as 6 × 7, 6 × 6, 4 × 7 and 4 × 8.

Let's see what happens when we try 6 × 7 using the method from Chapter 1.

We draw the circles below the numbers and subtract the numbers we are multiplying from 10. We write 4 and 3 in the circles. Our problem looks like this:

$$6 \quad \times \quad 7 \quad =$$

$$\text{(-4)} \qquad \text{(-3)}$$

Now we subtract crossways: 3 from 6 or 4 from 7 is 3. We write 3 after the equals sign.

$$6 \quad \times \quad 7 \quad = \quad \mathbf{3}$$

$$\text{(-4)} \qquad \text{(-3)}$$

Four times 3 is 12, so we write 12 after the 3 for an answer of 312.

$$6 \quad \times \quad 7 \quad = \quad 3\mathbf{12}$$

$$\text{(-4)} \qquad \text{(-3)}$$

Is this the correct answer? No, obviously it isn't.

Let's do the calculation again, this time using the reference number.

$$6 \quad \times \quad 7 \quad = \quad 30$$

$$-4 \qquad -3 \qquad + \underline{12}$$

$$42 \quad \textit{Answer}$$

That's more like it.

You should set out the calculations as shown above until the method is familiar to you, then you can simply use the reference number in your head.

Test yourself

Try these problems using a reference number of 10:

(a) 6 × 7 =   (d) 8 × 4 =

(b) 7 × 5 =   (e) 3 × 8 =

(c) 8 × 5 =   (f) 6 × 5 =

The answers are:

(a) 42   (b) 35   (c) 40
(d) 32   (e) 24   (f) 30

Using 100 as a reference number

What was our reference number for 96 × 97 in Chapter 1? One hundred, because we asked how many more do we need to make 100.

The problem worked out in full would look like this:

$$(100) \quad 96 \quad \times \quad 97 \quad = \quad 9{,}300$$
$$(-4) \qquad (-3) \qquad + \underline{\quad 12\quad}$$
$$9{,}312 \quad \textit{Answer}$$

The technique I explained for doing the calculations in your head actually makes you use this method. Let's multiply 98 by 98 and you will see what I mean.

If you take 98 and 98 from 100 you get answers of 2 and 2. Then take 2 from 98, which gives an answer of 96. If you were saying the answer aloud, you would not say, 'Ninety-six', you would say, 'Nine thousand, six hundred and ...'. Nine thousand, six hundred is the answer you get when you multiply 96 by the reference number, 100.

Now multiply the numbers in the circles: 2 times 2 is 4. You can now say the full answer: 'Nine thousand, six hundred and four.' Without using the reference number we might have just written the 4 after 96.

Here is how the calculation looks written in full:

$$(100) \quad 98 \quad \times \quad 98 \quad = \quad 9{,}600$$
$$(-2) \qquad (-2) \qquad + \underline{\quad 4\quad}$$
$$9{,}604 \quad \textit{Answer}$$

Test yourself

Do these problems in your head:

(a) 96 × 96 =

(b) 97 × 97 =

(c) 99 × 99 =

(d) 95 × 95 =

(e) 98 × 97 =

Your answers should be:

(a) 9,216   (b) 9,409   (c) 9,801
(d) 9,025   (e) 9,506

Double multiplication

What happens if you don't know your tables very well? How would you multiply 92 times 94? As we have seen, you would draw the circles below the numbers and write 8 and 6 in the circles. But if you don't know the answer to 8 times 6 you still have a problem.

You can get around this by combining the methods. Let's try it.

We write the problem and draw the circles:

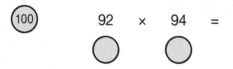

We write 8 and 6 in the circles.

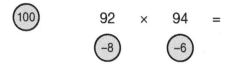

We subtract (take away) crossways: either 92 minus 6 or 94 minus 8.

I would choose 94 minus 8 because it is easy to subtract 8. The easy way to take 8 from a number is to take 10 and then add 2. Ninety-four minus 10 is 84, plus 2 is 86. We write 86 after the equals sign.

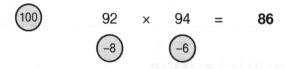

Now multiply 86 by the reference number, 100, to get 8,600. Then we must multiply the numbers in the circles: 8 times 6.

If we don't know the answer, we can draw two more circles below 8 and 6 and make another calculation. We subtract the 8 and 6 from 10, giving us 2 and 4. We write 2 in the circle below the 8, and 4 in the circle below the 6.

The calculation now looks like this:

(100)   92 × 94 = 8,600

         (-8)   (-6)

         (-2)   (-4)

We now need to calculate 8 times 6, using our usual method of subtracting diagonally. Two from 6 is 4, which becomes the first digit of this part of our answer.

We then multiply the numbers in the circles. This is 2 times 4, which is 8, the final digit. This gives us 48.

It is easy to add 8,600 and 48.

$8,600 + 48 = 8,648$

Here is the calculation in full.

(100)   92 × 94 = 8,600

         (-8)   (-6)   + ___48

         (-2)   (-4)   8,648 *Answer*

You can also use the numbers in the bottom circles to help your subtraction. The easy way to take 8 from 94 is to take 10 from 94, which is 84, and add the 2 in the circle to get 86. Or you could take 6 from 92. To do this, take 10 from 92, which is 82, and add the 4 in the circle to get 86.

With a little practice, you can do these calculations entirely in your head.

Note to parents and teachers

People often ask me, 'Don't you believe in teaching multiplication tables to children?'

My answer is, 'Yes, certainly I do. This method is the easiest way to teach the tables. It is the fastest way, the most painless way and the most pleasant way to learn tables.'

And while they are learning their tables, they are also learning basic number facts, practising addition and subtraction, memorising combinations of numbers that add to 10, working with positive and negative numbers, and learning a whole approach to basic mathematics.

Chapter 3
Numbers above the reference number

What if you want to multiply numbers above the reference number; above 10 or 100? Does the method still work? Let's find out.

Multiplying numbers in the teens

Here is how we multiply numbers in the teens. We will use 13 × 15 as an example and use 10 as our reference number.

           13   ×   15   =

Both 13 and 15 are above the reference number, 10, so we draw the circles above the numbers, instead of below as we have been doing. How much above 10 are they? Three and 5, so we write 3 and 5 in the circles above 13 and 15. Thirteen is 10 plus 3, so we write a plus sign in front of the 3; 15 is 10 plus 5, so we write a plus sign in front of the 5.

As before, we now go crossways. Thirteen plus 5 or 15 plus 3 is 18. We write 18 after the equals sign.

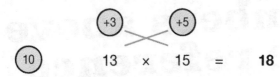

We then multiply the 18 by the reference number, 10, and get 180. (To multiply a number by 10 we add a 0 to the end of the number.) One hundred and eighty is our subtotal, so we write 180 after the equals sign.

(+3)   (+5)

(10)   13   ×   15   =   180

For the last step, we multiply the numbers in the circles. Three times 5 equals 15. Add 15 to 180 and we get our answer of 195. This is how we write the problem in full:

(+3) ——— (+5)

(10)   13   ×   15   =   180

+ <u>15</u>

195   *Answer*

*If the number we are multiplying is **above** the reference number we put the circle **above**. If the number is **below** the reference number we put the circle **below**.*

*If the circled number is **above** we **add** diagonally.*

*If the circled number is **below** we **subtract** diagonally.*

*The numbers in the circles **above** are **plus** numbers and the numbers in the circles **below** are **minus** numbers.*

Let's try another one. How about 12 × 17?

The numbers are above 10 so we draw the circles above. How much above 10? Two and 7, so we write 2 and 7 in the circles.

$$\begin{array}{ccccc} & \boxed{+2} & & \boxed{+7} & \\ \boxed{10} & 12 & \times & 17 & = \end{array}$$

What do we do now? Because the circles are above, the numbers are plus numbers so we add crossways. We can either do 12 plus 7 or 17 plus 2. Let's do 17 plus 2.

$17 + 2 = 19$

We now multiply 19 by 10 (our reference number) to get 190 (we just put a 0 after the 19). Our work now looks like this:

$$\begin{array}{ccccc} & \boxed{+2} & & \boxed{+7} & \\ \boxed{10} & 12 & \times & 17 & = & \mathbf{190} \end{array}$$

Now we multiply the numbers in the circles.

$2 \times 7 = 14$

Add 14 to 190 and we have our answer. Fourteen is 10 plus 4. We can add the 10 first (190 + 10 = 200), then the 4, to get 204.

Here is the finished problem:

(+2)   (+7)

(10)   12   ×   17   =   190
                              + _14_
                              204   *Answer*

Test yourself

Now try these problems by yourself.

(a) 12 × 15 =   (f) 12 × 16 =

(b) 13 × 14 =   (g) 14 × 14 =

(c) 12 × 12 =   (h) 15 × 15 =

(d) 13 × 13 =   (i) 12 × 18 =

(e) 12 × 14 =   (j) 16 × 14 =

The answers are:

(a) 180	(b) 182	(c) 144	(d) 169
(e) 168	(f) 192	(g) 196	(h) 225
(i) 216	(j) 224		

If any of your answers were wrong, read through this section again, find your mistake, then try again.

How would you solve 13 × 21? Let's try it:

(10) 13 × 21 =

We still use a reference number of 10. Both numbers are above 10 so we put the circles above. Thirteen is 3 above 10, 21 is 11 above, so we write 3 and 11 in the circles.

Twenty-one plus 3 is 24, times 10 is 240. Three times 11 is 33, added to 240 makes 273. This is how the completed problem looks:

```
       (+3)        (+11)
(10)    13   ×   21   =   240
                       +   33
                          273   Answer
```

Multiplying numbers above 100

We can use our speed maths method to multiply numbers above 100 as well. Let's try 113 times 102.

We use 100 as our reference number.

```
       (+13)       (+2)
(100)   113   ×   102   =
```

Add crossways:

113 + 2 = 115

25

Multiply by the reference number:

$115 \times 100 = 11{,}500$

Now multiply the numbers in the circles:

$2 \times 13 = 26$

This is how the completed problem looks:

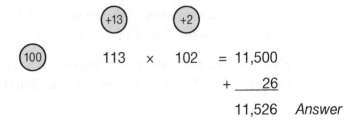

$$113 \quad \times \quad 102 \quad = 11{,}500$$
$$+ \quad \underline{ 26}$$
$$11{,}526 \quad \textit{Answer}$$

Solving problems in your head

When you use these strategies, what you say inside your head is very important, and can help you solve problems more quickly and easily.

Let's try multiplying 16 by 16.

This is how I would solve this problem in my head:

> 16 plus 6 (from the second 16) equals 22, times 10 equals 220

> 6 times 6 is 36

> 220 plus 30 is 250, plus 6 is 256

Try it. See how you go.

Inside your head you would say:

> 16 plus 6 … 22 … 220 … 36 … 256

With practice, you can leave out a lot of that. You don't have to go through it step by step. You would only say to yourself:

200 ... 256

Practise doing this. Saying the right thing in your head as you do the calculation can better than halve the time it takes.

How would you calculate 7 × 8 in your head? You would 'see' 3 and 2 below the 7 and 8. You would take 2 from the 7 (or 3 from the 8) and say, 'Fifty', multiplying by 10 in the same step. Three times 2 is 6. All you would say is, 'Fifty ... six.'

What about 6 × 7?

You would 'see' 4 and 3 below the 6 and 7. Six minus 3 is 3; you say, 'Thirty'. Four times 3 is 12, plus 30 is 42. You would just say, 'Thirty ... forty-two.'

It's not as hard as it sounds, is it? And it will become easier the more you do.

Double multiplication

Let's multiply 88 by 84. We use 100 as our reference number. Both numbers are below 100 so we draw the circles below. How many below are they? Twelve and 16. We write 12 and 16 in the circles. Now subtract crossways: 84 minus 12 is 72.

(Subtract 10, then 2, to subtract 12.)

Multiply the answer of 72 by the reference number, 100, to get 7,200.

The calculation so far looks like this:

(100)　　　　88　×　　84　　=　7,200

　　　(−12)　　　(−16)

We now multiply 12 times 16 to finish the calculation.

　　　(+2)　　　(+6)

(10)　　　　12　×　　16　　=　　　180

　　　　　　　　　　　　　　+　 12

　　　　　　　　　　　　　　　　192

This calculation can be done mentally.

Now add this answer to our subtotal of 7,200.

If you were doing the calculation in your head you would simply add 100 first, then 92, like this: 7,200 plus 100 is 7,300, plus 92 is 7,392. Simple.

You should easily do this in your head with just a little practice.

Test yourself

Try these problems:

(a)　87 × 86 =

(c)　88 × 87 =

(b)　88 × 88 =

(d)　88 × 85 =

The answers are:

(a) 7,482   (b) 7,744   (c) 7,656   (d) 7,480

Combining the methods taught in this book creates endless possibilities. Experiment for yourself.

Note to parents and teachers

This chapter introduces the concept of positive and negative numbers. We will simply refer to them as plus and minus numbers throughout the book.

These methods make positive and negative numbers tangible. Children can easily relate to the concept because it is made visual.

Calculating numbers in the eighties using double multiplication develops concentration. I find most children can do the calculations much more easily than most adults think they should be able to.

Kids love showing off. Give them the opportunity.

Chapter 4
Multiplying above & below the reference number

Until now, we have multiplied numbers that were both below the reference number or both above the reference number. How do we multiply numbers when one number is above the reference number and the other is below the reference number?

Numbers above and below

We will see how this works by multiplying 97 × 125. We will use 100 as our reference number:

   97   ×   125   =

Ninety-seven is below the reference number, 100, so we put the circle below. How much below? Three, so we write 3 in the circle. One hundred and twenty-five is above so we put the circle above. How much above? Twenty-five, so we write 25 in the circle above.

$$
\begin{array}{c}
\quad\quad\quad\quad (+25) \\
(100) \quad\quad 97 \;\times\; 125 \;=\; \\
\quad\quad\quad (-3)
\end{array}
$$

One hundred and twenty-five is 100 plus 25 so we put a plus sign in front of the 25. Ninety-seven is 100 minus 3 so we put a minus sign in front of the 3.

We now calculate crossways. Either 97 plus 25 or 125 minus 3. One hundred and twenty-five minus 3 is 122. We write 122 after the equals sign. We now multiply 122 by the reference number, 100. One hundred and twenty-two times 100 is 12,200. (To multiply any number by 100, we simply put two zeros after the number.) This is similar to what we have done in earlier chapters.

This is how the problem looks so far:

$$
\begin{array}{c}
\quad\quad\quad\quad (+25) \\
(100) \quad\quad 97 \;\times\; 125 \;=\; 12{,}200 \\
\quad\quad\quad (-3)
\end{array}
$$

Now we multiply the numbers in the circles. Three times 25 is 75, but that is not really the problem. We have to multiply 25 by minus 3. The answer is –75.

Now our problem looks like this:

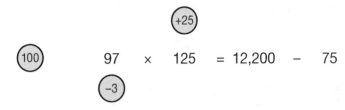

$$97 \quad \times \quad 125 \quad = 12,200 \quad - \quad 75$$

A short cut for subtraction

Let's take a break from this problem for a moment to have a look at a short cut for the subtractions we are doing.

What is the easiest way to subtract 75? Let me ask another question. What is the easiest way to take 9 from 63 in your head?

$63 - 9 =$

I am sure you got the right answer, but how did you get it? Some would take 3 from 63 to get 60, then take another 6 to make up the 9 they have to take away, and get 54.

Some would take away 10 from 63 and get 53. Then they would add 1 back because they took away 1 too many. This would also give 54.

Some would do the problem the same way they would when using pencil and paper. This way they have to carry and borrow in their heads. This is probably the most difficult way to solve the problem.

Remember, the easiest way to solve a problem is also the fastest, with the least chance of making a mistake.

Most people find the easiest way to subtract 9 is to take away 10, then add 1 to the answer. The easiest way to subtract 8 is to take away 10, then add 2 to the answer. The easiest way to subtract 7 is to take away 10, then add 3 to the answer.

What is the easiest way to take 90 from a number? Take 100 and give back 10.

What is the easiest way to take 80 from a number? Take 100 and give back 20.

What is the easiest way to take 70 from a number? Take 100 and give back 30.

If we go back to the problem we were working on, how do we take 75 from 12,200? We can take away 100 and give back 25. Is this easy? Let's try it. Twelve thousand, two hundred minus 100? Twelve thousand, one hundred. Plus 25? Twelve thousand, one hundred and twenty-five.

So back to our example. This is how the completed problem looks:

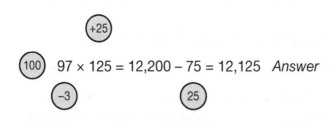

With a little practice you should be able to solve these problems entirely in your head. Practise with the following problems.

Teach yourself

Try these

a)   98 × 145 =

e)   98 × 146 =

b)   98 × 125 =

f)   9 × 15 =

c)   95 × 120 =

g)   8 × 12 =

d)   96 × 125 =

h)   7 × 12 =

How did you go? The answers are:

a) 14,210   b) 12,250   c) 11,400   d) 12,000
e) 14,308   f) 135   g) 96   h) 84

Multiplying numbers in the circles

The rule for multiplying the numbers in the circles is:

> *When **both** circles are above the numbers or **both** circles are below the numbers, we **add** the answer. When **one** circle is above and **one** circle is below we **subtract**.*

Mathematically, we would say: when we multiply two positive (plus) numbers we get a positive (plus) answer. When we multiply two negative (minus) numbers we get a positive (plus) answer. When we multiply a positive (plus) by a negative (minus) we get a minus answer.

Let's try another problem. Would our method work for multiplying 8 × 42? Let's try it.

We choose a reference number of 10. Eight is 2 below 10 and 42 is 32 above 10.

$$\text{(10)} \quad 8 \;\times\; 42 \;=\;$$

(+32)

(−2)

We either take 2 from 42 or add 32 to 8. Two from 42 is 40, times the reference number, 10, is 400. Minus 2 times 32 is −64. To take 64 from 400 we take 100, which equals 300, then give back 36 for a final answer of 336. (We will look at an easy way to subtract numbers from 100 in the chapter on subtraction.)

Our completed problem looks like this:

(+32)

(10) 8 × 42 = 400 − 64 = 336 *Answer*

(−2)   (36)

We haven't finished with multiplication yet, but we can take a rest here and practise what we have already learnt. If some problems don't seem to work out easily, don't worry; we still have more to cover.

In the next chapter we will have a look at a simple method for checking answers.

Chapter 5
Checking your answers

What would it be like if you always found the right answer to every maths problem? Imagine scoring 100% for every maths test. How would you like to get a reputation for never making a mistake? If you do make a mistake, I can teach you how to find and correct it before anyone (including your teacher) knows anything about it.

When I was young, I often made mistakes in my calculations. I knew how to do the problems, but I still got the wrong answer. I would forget to carry a number, or find the right answer but write down something different, and who knows what other mistakes I would make.

I had some simple methods for checking answers I had devised myself but they weren't very good. They would confirm maybe the last digit of the answer or they would show me the answer I got was at least close to the real answer. I wish I had known then the method I am going to show you now. Everyone would have thought I was a genius if I had known this.

Mathematicians have known this method of checking answers for about 1,000 years, although I have made a small change I haven't seen anywhere else. It is called the digit sum method. I have taught this method of checking answers in my other books, but this time I am going to teach it differently. This method of checking your answers will work for almost any calculation. Because I still make mistakes occasionally, I always check my answers. Here is the method I use.

Substitute numbers

To check the answer to a calculation, we use substitute numbers instead of the original numbers we were working with. A substitute on a football team or a basketball team is somebody who takes another person's place on the team. If somebody gets injured, or tired, they take that person off and bring on a substitute player. A substitute teacher fills in when your regular teacher is unable to teach you. We can use substitute numbers in place of the original numbers to check our work. The substitute numbers are always low and easy to work with.

Let me show you how it works. Let us say we have just calculated 12×14 and come to an answer of 168. We want to check this answer.

$12 \times 14 = 168$

The first number in our problem is 12. We add its digits together to get the substitute:

$1 + 2 = 3$

Three is our substitute for 12. I write 3 in pencil either above or below the 12, wherever there is room.

The next number we are working with is 14. We add its digits:

$1+4=5$

Five is our substitute for 14.

We now do the same calculation (multiplication) using the substitute numbers instead of the original numbers:

$3\times5=15$

Fifteen is a two-digit number so we add its digits together to get our check answer:

$1+5=6$

Six is our check answer.

We add the digits of the original answer, 168:

$1+6+8=15$

Fifteen is a two-digit number so we add its digits together to get a one-digit answer:

$1+5=6$

Six is our substitute answer. This is the same as our check answer, so our original answer is correct.

Had we got an answer that added to, say, 2 or 5, we would know we had made a mistake. The substitute answer must be the same as the check answer if the substitute is correct. If our substitute answer is different we know we have to go back and check our work to find the mistake.

I write the substitute numbers in pencil so I can erase them when I have made the check. I write the substitute numbers either above or below the original numbers, wherever I have room.

The example we have just done would look like this:

$$\overset{\textcircled{+2}}{12} \quad \times \quad \overset{\textcircled{+4}}{14} \quad = \quad 160$$

$$\textcircled{10} \qquad 3 \qquad\qquad 5 \qquad\qquad + \underline{8}$$

$$168$$

$$6$$

If we have the right answer in our calculation, the digits in the original answer should add up to the same as the digits in our check answer.

Let's try it again, this time using 14 × 14:

$14 \times 14 = 196$

$1 + 4 = 5$ (substitute for 14)

$1 + 4 = 5$ (substitute for 14 again)

So our substitute numbers are 5 and 5. Our next step is to multiply these:

$5 \times 5 = 25$

Twenty-five is a two-digit number so we add its digits:

$2 + 5 = 7$

Seven is our check answer.

Now, to find out if we have the correct answer, we add the digits in our original answer, 196:

$1 + 9 + 6 = 16$

To bring 16 to a one-digit number:

$1 + 6 = 7$

Seven is what we got for our check answer so we can be confident we didn't make a mistake.

$$
\begin{array}{ccccc}
& \textcircled{+4} & & \textcircled{+4} & \\
\textcircled{10} & 14 & \times & 14 & = \quad 180 \\
& 5 & & 5 & + \ \underline{16} \\
& & & & 196 \\
& & & & 7
\end{array}
$$

A short cut

There is another short cut to this procedure. If we find a 9 anywhere in the calculation, we cross it out. This is called casting out nines. You can see with this example how this removes a step from our calculations without affecting the result. With the last answer, 196, instead of adding 1 + 9 + 6, which equals 16, and then adding 1 + 6, which equals 7, we could cross out the 9 and just add 1 and 6, which also equals 7. This makes no difference to the answer, but it saves some time and effort, and I am in favour of anything that saves time and effort.

What about the answer to the first problem we solved, 168? Can we use this short cut? There isn't a 9 in 168.

We added 1 + 6 + 8 to get 15, then added 1 + 5 to get our final check answer of 6. In 168, we have two digits that add up to 9, the 1 and the 8. Cross them out and you just have the 6 left. No more work to do at all, so our short cut works.

Check any size number

What makes this method so easy to use is that it changes any size number into a single-digit number. You can check calculations that are too big to go into your calculator by casting out nines.

For instance, if we wanted to check 12,345,678 × 89,045 = 1,099,320,897,510, we would have a problem because most calculators can't handle the number of digits in the answer, so most would show the first digits of the answer with an error sign.

The easy way to check the answer is to cast out the nines. Let's try it.

~~12,345,678~~   0
~~89,045~~ =   8
~~1,099,320,897,510~~   0

All of the digits in the answer cancel. The nines automatically cancel, then we have 1 + 8, 2 + 7, then 3 + 5 + 1 = 9, which cancels again. And 0 × 8 = 0, so our answer seems to be correct.

Let's try it again.

$$137 \times 456 = 62,472$$

To find our substitute for 137:

$$1 + 3 + 7 = 11$$
$$1 + 1 = 2$$

There were no short cuts with the first number. Two is our substitute for 137.

To find our substitute for 456:

$4 + 5 + 6 =$

We immediately see that $4 + 5 = 9$, so we cross out the 4 and the 5. That just leaves us with 6, our substitute for 456.

Can we find any nines, or digits adding up to 9, in the answer? Yes, $7 + 2 = 9$, so we cross out the 7 and the 2. We add the other digits:

$6 + 2 + 4 = 12$
$1 + 2 + 3$

Three is our substitute answer.

I write the substitute numbers in pencil above or below the actual numbers in the problem. It might look like this:

$$137 \times \cancel{456} = 62,4\cancel{72}$$
$$\quad 2 \qquad 6 \qquad\quad 3$$

Is 62,472 the right answer?

We multiply the substitute numbers: 2 times 6 equals 12. The digits in 12 add up to 3 ($1 + 2 = 3$). This is the same as our substitute answer, so we were right again.

Let's try one more example. Let's check if this answer is correct:

$$456 \times 831 = 368,936$$

We write in our substitute numbers:

$$\cancel{456} \times \cancel{831} = \cancel{368},\cancel{936}$$
$$\quad 6 \qquad\quad 3 \qquad\quad 8$$

That was easy because we cast out (or crossed out) 4 and 5 from the first number, leaving 6. We cast out 8 and 1 from the second number, leaving 3. And almost every digit was cast out of the answer, 3 plus 6 twice, and a 9, leaving a substitute answer of 8.

We now see if the substitutes work out correctly: 6 times 3 is 18, which adds up to 9, which also gets cast out, leaving 0. But our substitute answer is 8, so we have made a mistake somewhere.

When we calculate it again, we get 378,936.

Did we get it right this time? The 936 cancels out, so we add 3 + 7 + 8, which equals 18, and 1 + 8 adds up to 9, which cancels, leaving 0.

This is the same as our check answer, so this time we have it right.

Does this method prove we have the right answer? No, but we can be almost certain.

This method won't find *all* mistakes. For instance, say we had 3,789,360 for our last answer; by mistake we put a 0 on the end. The final 0 wouldn't affect our check by casting out nines and we wouldn't know we had made a mistake. When it showed we had made a mistake, though, the check definitely proved we had the wrong answer. It is a simple, fast check that will find most mistakes, and should get you 100% scores in most of your maths tests.

Do you get the idea? If you are unsure about using this method to check your answers, we will be using the method throughout the book so you will soon become familiar with it. Try it on your calculations at school and at home.

Why does the method work?

You will be much more successful using a new method when you know not only that it does work, but you understand why it works as well.

Firstly, 10 is 1 times 9 with 1 remainder. Twenty is 2 nines with 2 remainder. Twenty-two would be 2 nines with 2 remainder for the 20 plus 2 more for the units digit.

If you have 35 cents in your pocket and you want to buy as many lollies as you can for 9 cents each, each 10 cents will buy you one lolly with 1 cent change. So, 30 cents will buy you three lollies with 3 cents change, plus the extra 5 cents in your pocket gives you 8 cents. So, the number of tens plus the units digit gives you the nines remainder.

Secondly, think of a number and multiply it by 9. What is 4 × 9? The answer is 36. Add the digits in the answer together, 3 + 6, and you get 9.

Let's try another number. Three nines are 27. Add the digits of the answer together, 2 + 7, and you get 9 again.

Eleven nines are 99. Nine plus 9 equals 18. Wrong answer? No, not yet. Eighteen is a two-digit number so we add its digits together: 1 + 8. Again, the answer is 9.

If you multiply any number by 9, the sum of the digits in the answer will always add up to 9 if you keep adding the digits until you get a one-digit number. This is an easy way to tell if a number is evenly divisible by 9. If the digits of any number add up to 9, or a multiple of 9, then the number itself is evenly divisible by 9.

If the digits of a number add up to any number other than 9, this other number is the remainder you would get after dividing the number by 9.

Let's try 13:

  1+3 = 4

Four is the digit sum of 13. It should be the remainder you would get if you divided by 9. Nine divides into 13 once, with 4 remainder.

If you add 3 to the number, you add 3 to the remainder. If you double the number, you double the remainder. If you halve the number you halve the remainder.

Don't believe me? Half of 13 is 6.5. Six plus 5 equals 11. One plus 1 equals 2. Two is half of 4, the nines remainder for 13.

Whatever you do to the number, you do to the remainder, so we can use the remainders as substitutes.

Why do we use 9 remainders? Couldn't we use the remainders after dividing by, say, 17? Certainly, but there is so much work involved in dividing by 17, the check would be harder than the original problem. We choose 9 because of the easy short cut method for finding the remainder.

Chapter 6
Multiplication using any reference number

In Chapters 1 to 4 you learnt how to multiply numbers using an easy method that makes multiplication fun. It is easy to use when the numbers are near 10 or 100. But what about multiplying numbers that are around 30 or 60? Can we still use this method? We certainly can.

We chose reference numbers of 10 and 100 because it is easy to multiply by 10 and 100. The method will work just as well with other reference numbers, but we must choose numbers that are easy to multiply by.

Multiplication by factors

It is easy to multiply by 20, because 20 is 2 times 10. It is easy to multiply by 10 and it is easy to multiply by 2. This is called multiplication by factors, because 10 and 2 are factors of 20 (20 = 10 × 2). So, to multiply any number by 20, you multiply it by 2 and then multiply the answer by 10, or, you could say, you double the number and add a 0.

For instance, to multiply 7 by 20 you would double it (2 × 7 = 14) then multiply your answer by 10 (14 × 10 = 140). To multiply 32 by 20 you would double 32 (64) then multiply by 10 (640).

Multiplying by 20 is easy because it is easy to multiply by 2 and it is easy to multiply by 10.

So, it is easy to use 20 as a reference number. Let us try an example:

23 × 21 =

Twenty-three and 21 are just above 20, so we use 20 as our reference number. Both numbers are above 20 so we put the circles above. How much above are they? Three and 1. We write those numbers above in the circles. Because the circles are above they are plus numbers.

We add diagonally:

23 + 1 = 24

We multiply the answer, 24, by the reference number, 20. To do this we multiply by 2, then by 10:

24 × 2 = 48
48 × 10 = 480

We could now draw a line through the 24 to show we have finished using it.

The rest is the same as before. We multiply the numbers in the circles:

$3 \times 1 = 3$
$480 + 3 = 483$

The problem now looks like this:

$$
\begin{array}{ccccc}
& & & & \boxed{+3} \qquad \boxed{+1} \\
\boxed{20} & & 23 & \times & 21 & = & \cancel{24} \\
& & & & & & 480 \\
& & & & & + & \underline{3} \\
& & & & & & 483 \quad \textit{Answer}
\end{array}
$$

Checking answers

Let us apply what we learnt in the last chapter and check our answer:

$$
\begin{array}{ccccc}
23 & \times & 21 & = & 483 \\
5 & & 3 & & 15 \\
& & & & 6
\end{array}
$$

The substitute numbers for 23 and 21 are 5 and 3.

$5 \times 3 = 15$
$1 + 5 = 6$

Six is our check answer.

The digits in our original answer, 483, add up to 6:

$4 + 8 + 3 = 15$
$1 + 5 = 6$

This is the same as our check answer, so we were right. Let's try another:

$23 \times 31 =$

We put 3 and 11 above 23 and 31:

They are 3 and 11 above the reference number, 20. Adding diagonally, we get 34:

$31 + 3 = 34$

or

$23 + 11 = 34$

We multiply this answer by the reference number, 20. To do this, we multiply 34 by 2, then multiply by 10.

$34 \times 2 = 68$
$68 \times 10 = 680$

This is our subtotal. We now multiply the numbers in the circles (3 and 11):

$3 \times 11 = 33$

Add this to 680:

$680 + 33 = 713$

The calculation will look like this:

$$\overset{(20)}{} \quad \overset{(+3)}{23} \quad \times \quad \overset{(+11)}{31} \quad = \quad \cancel{34}$$

$$680$$
$$+ \ \underline{\ 33}$$
$$713 \quad \textit{Answer}$$

We check by using substitute numbers:

$$23 \ \times \ 31 \ = \ 713$$
$$\ 5 \qquad 4 \qquad 11$$
$$\qquad\qquad\qquad 2$$

Multiply our substitute numbers and then add the digits in the answer:

$5 \times 4 = 20$

$2 + 0 = 2$

This checks with our substitute answer so we can accept that as correct.

Test yourself

Here are some problems to try. When you have finished them, you can check your answers yourself by casting out the nines.

(a) 21 × 24 =

(d) 23 × 27 =

(b) 24 × 24 =

(e) 21 × 35 =

(c) 23 × 23 =

(f) 26 × 24 =

You should be able to do all of those problems in your head. It's not difficult with a little practice.

Multiplying numbers below 20

How about multiplying numbers below 20? If the numbers (or one of the numbers to be multiplied) are in the high teens, we can use 20 as a reference number.

Let's try an example:

$18 \times 17 =$

Using 20 as a reference number we get:

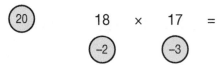

Subtract diagonally:

$17 - 2 = 15$

Multiply by 20:

$2 \times 15 = 30$
$30 \times 10 = 300$

Three hundred is our subtotal.

Now we multiply the numbers in the circles and then add the result to our subtotal:

$2 \times 3 = 6$
$300 + 6 = 306$

Our completed work should look like this:

(20) 18   ×   17   =   ~~15~~

       (-2) (-3) 300

                           +   <u> 6</u>

                          306   *Answer*

Now let's try the same example using 10 as a reference number:

       $(+8)$ $(+7)$

(10) 18   ×   17   =

Add crossways, then multiply by 10 to get a subtotal:

$18 + 7 = 25$
$10 \times 25 = 250$

Multiply the numbers in the circles and add this to the subtotal:

$8 \times 7 = 56$
$250 + 56 = 306$

Our completed work should look like this:

       $(+8)$ $(+7)$

(10) 18   ×   17   =   250

                           + <u>56</u>

                          306   *Answer*

This confirms our first answer. There isn't much to choose between using the different reference numbers. It is a matter of personal preference. Simply choose the reference number you find easier to work with.

Multiplying numbers above and below 20

The third possibility is if one number is above and the other below 20.

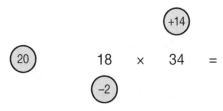

We can either add 18 to 14 or subtract 2 from 34, and then multiply the result by our reference number:

34 − 2 = 32
32 × 20 = 640

We now multiply the numbers in the circles:

2 × 14 = 28

It is actually −2 times 14 so our answer is −28.

640 − 28 = 612

To subtract 28, we subtract 30 and add 2.

640 − 30 = 610
610 + 2 = 612

Our problem now looks like this:

(+14)

(20) 18 × 34 = ~~32~~

(−2)   640 − 28 = 612   *Answer*

(2)

Let's check the answer by casting out the nines:

$$18 \times 34 = 612$$
$$9 \quad\quad 7 \quad\quad 9$$
$$0 \quad\quad\quad\quad\quad 0$$

Zero times 7 is 0, so the answer is correct.

Using 50 as a reference number

That takes care of the numbers up to around 30 times 30. What if the numbers are higher? Then we can use 50 as a reference number. It is easy to multiply by 50 because 50 is half of 100. You could say 50 is 100 divided by 2. So, to multiply by 50, we multiply the number by 100 and then divide the answer by 2.

Let's try it:

$$(50) \quad\quad 47 \quad \times \quad 45 \quad =$$
$$(-3) \quad\quad\quad (-5)$$

Subtract diagonally:

$$45 - 3 = 42$$

Multiply 42 by 100, then divide by 2:

$$42 \times 100 = 4,200$$
$$4,200 \div 2 = 2,100$$

Now multiply the numbers in the circles, and add this result to 2,100:

$$3 \times 5 = 15$$
$$2,100 + 15 = 2,115$$

(50)   47   ×   45   =   ~~4,200~~

(-3)   (-5)   2,100

                        +   15

                            2,115   *Answer*

Fantastic. That was so easy. Let's try another:

           (+3)   (+7)

(50)   53   ×   57   =

Add diagonally, then multiply the result by the reference number (multiply by 100 and then divide by 2):

$57 + 3 = 60$
$60 \times 100 = 6,000$

We divide by 2 to get 3,000. We then multiply the numbers in the circles and add the result to 3,000:

$3 \times 7 = 21$
$3,000 + 21 = 3,021$

Our problem will end up looking like this:

           (+3)   (+7)

(50)   53   ×   57   =   ~~6,000~~

                            3,000

                        +   21

                            3,021   *Answer*

Let's try one more:

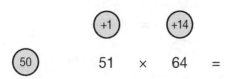

$$51 \quad \times \quad 64 \quad =$$

Add diagonally and multiply the result by the reference number (multiply by 100 and then divide by 2):

$64 + 1 = 65$
$65 \times 100 = 6,500$

Then we halve the answer.

Half of 6,000 is 3,000. Half of 500 is 250. Our subtotal is 3,250.

Now multiply the numbers in the circles:

$1 \times 14 = 14$

Add 14 to the subtotal to get 3,264.

Our problem now looks like this:

$$51 \quad \times \quad 64 \quad = \quad \cancel{6,500}$$
$$3,250$$
$$+ \quad \underline{14}$$
$$3,264 \quad \textit{Answer}$$

We could check that by casting out the nines:

$$51 \quad \times \quad 64 \quad = \quad 3,264$$
$$6 \qquad\quad 1 \qquad\qquad 6$$

Six and 4 in 64 add up to 10, which added again gives us 1.

Six times 1 does give us 6, so the answer is correct.

Test yourself

Here are some problems for you to do. See how many you can do in your head.

(a) 45 × 45 =

(e) 51 × 57 =

(b) 49 × 49 =

(f) 54 × 56 =

(c) 47 × 43 =

(g) 51 × 68 =

(d) 44 × 44 =

(h) 51 × 72 =

How did you go with those? You should have had no trouble doing all of them in your head. The answers are:

| (a) 2,025 | (b) 2,401 | (c) 2,021 | (d) 1,936 |
| (e) 2,907 | (f) 3,024 | (g) 3,468 | (h) 3,672 |

Multiplying higher numbers

There is no reason why we can't use 200, 500 and 1,000 as reference numbers.

To multiply by 200 we multiply by 2 and 100. To multiply by 500 we multiply by 1,000 and halve the answer. To multiply by 1,000 we simply write three zeros after the number.

Let's try some examples.

212 × 212 =

We use 200 as a reference number.

Both numbers are above the reference number so we draw the circles above the numbers. How much above? Twelve and 12, so we write 12 in each circle.

They are plus numbers so we add crossways.

212 + 12 = 224

We multiply 224 by 2 and by 100.

224 × 2 = 448
448 × 100 = 44,800

Now multiply the numbers in the circles.

12 × 12 = 144

(You must know 12 times 12. If you don't, you simply calculate the answer using 10 as a reference number.)

44,800 + 144 = 44,944

(+12)   (+12)

(200)   212 × 212 = ~~224~~

44,800

+   144

44,944 *Answer*

Let's try another. How about multiplying 511 by 503? We use 500 as a reference number.

$$511 \times 503 =$$

Add crossways.

$$511 + 3 = 514$$

Multiply by 500. (Multiply by 1,000 and divide by 2.)

$$514 \times 1,000 = 514,000$$

Half of 514,000 is 257,000. (How did we work this out? Half of 500 is 250 and half of 14 is 7.)

Multiply the numbers in the circles, and add the result to our subtotal.

$$3 \times 11 = 33$$
$$257,000 + 33 = 257,033$$

$$511 \times 503 = \cancel{514}$$

$$
\begin{array}{r}
257,000 \\
+ 33 \\
\hline
257,033 \quad \textit{Answer}
\end{array}
$$

Let's try one more.

$$989 \times 994 =$$

We use 1,000 as the reference number.

(1,000)   989 × 994 =

(−11)   (−6)

Subtract crossways.

989 − 6 = 983

or

994 − 11 = 983

Multiply our answer by 1,000.

983 × 1,000 = 983,000

Multiply the numbers in the circles.

11 × 6 = 66

983,000 + 66 = 983,066

(1,000)   989 × 994 = 983,000

(−11)   (−6) + _____66

983,066   *Answer*

Doubling and halving numbers

To use 20 and 50 as reference numbers, we need to be able to double and halve numbers easily.

Sometimes, like when you halve a two-digit number and the tens digit is odd, the calculation is not so easy.

For example:

78 ÷ 2 =

To halve 78, you might halve 70 to get 35, then halve 8 to get 4, and add the answers, but there is an easier method.

Seventy-eight is 80 minus 2. Half of 80 is 40 and half of −2 is −1. So, half of 78 is 40 minus 1, or 39.

Test yourself

Try these for yourself:

(a) 58 ÷ 2 =

(b) 76 ÷ 2 =

(c) 38 ÷ 2 =

(d) 94 ÷ 2 =

(e) 54 ÷ 2 =

(f) 36 ÷ 2 =

(g) 78 ÷ 2 =

(h) 56 ÷ 2 =

The answers are:

(a) 29	(b) 38	(c) 19	(d) 47
(e) 27	(f) 18	(g) 39	(h) 28

To double 38, think of 40 – 2. Double 40 would be 80 and double –2 is –4, so we get 80 – 4, which is 76. Again, this is useful when the units digit is high. With doubling, it doesn't matter if the tens digit is odd or even.

Test yourself

Now try these:

(a) 18 × 2 =

(b) 39 × 2 =

(c) 49 × 2 =

(d) 67 × 2 =

(e) 77 × 2 =

(f) 48 × 2 =

The answers are:

(a) 36	(b) 78	(c) 98
(d) 134	(e) 154	(f) 96

This strategy can easily be used to multiply or divide larger numbers by 3 or 4. For instance:

$$19 \times 3 = (20 - 1) \times 3 = 60 - 3 = 57$$
$$38 \times 4 = (40 - 2) \times 4 = 160 - 8 = 152$$

Note to parents and teachers

This strategy encourages the student to look at numbers differently. Traditionally, we have looked at a number like 38 as thirty, or three tens, plus eight ones. Now we are teaching students to also see the number 38 as 40 minus 2. Both concepts are correct.

Chapter 7
Multiplying lower numbers

You may have noticed that our method of multiplication doesn't seem to work with some numbers. For instance, let's try 6 × 4.

$$\textcircled{10} \qquad 6 \quad \times \quad 4 \quad =$$
$$\qquad\qquad \textcircled{-4} \qquad \textcircled{-6}$$

We use a reference number of 10. The circles go below because the numbers 6 and 4 are below 10. We subtract crossways, or diagonally.

$6 - 6 = 0$

or

$4 - 4 = 0$

We then multiply the numbers in the circles:

$4 \times 6 =$

That was our original problem. The method doesn't seem to help.

Is there a way to make the method work in this case? There is, but we must use a different reference number. The problem is not with the method but with the choice of reference number.

Let's try a reference number of 5. Five equals 10 divided by 2, or half of 10. The easy way to multiply by 5 is to multiply by 10 and halve the answer.

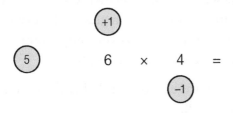

Six is <u>above</u> 5 so we put the circle above. Four is <u>below</u> 5 so we put the circle below. Six is 1 higher than 5 and 4 is 1 lower, so we put 1 in each circle.

We add or subtract crossways:

$6 - 1 = 5$

or

$4 + 1 = 5$

We multiply 5 by the reference number, which is also 5.

To do this, we multiply by 10, which gives us 50, and then divide by 2, which gives us 25. Now we multiply the numbers in the circles:

$1 \times -1 = -1$

Because the result is a negative number, we subtract it from our subtotal rather than add it:

25 − 1 = 24

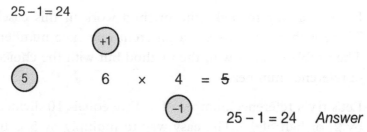

25 − 1 = 24   *Answer*

This is a long and complicated method for multiplying low numbers, but it shows we can make the method work with a little ingenuity. Actually, these strategies will develop your ability to think laterally, which is very important for mathematicians, and also for succeeding in life.

Let's try some more, even though you probably know your lower times tables quite well:

Subtract crossways:

4 − 1 = 3

Multiply your answer by the reference number:

3 × 10 = 30

Thirty divided by 2 equals 15. Now multiply the numbers in the circles:

1 × 1 = 1

Add that to our subtotal:

15 + 1 = 16

⑤   4   ×   4   =   ~~30~~
        ⓘ₋₁   ⓘ₋₁   15
                    + 1
                    16   *Answer*

Test yourself

Now try the following:

(a) 3 × 4 =   (d) 3 × 6 =

(b) 3 × 3 =   (e) 3 × 7 =

(c) 6 × 6 =   (f) 4 × 7 =

The answers are:

(a) 12   (b) 9   (c) 36
(d) 18   (e) 21   (f) 28

I'm sure you had no trouble doing those.

I don't really think this is the easiest way to learn those tables. I think it is easier to simply remember them. Some people want to learn how to multiply low numbers just to check that the method will work. Others like to know that if they can't remember some of their tables, there is an easy method to calculate the answer. Even if you know your tables for these numbers, it is still fun to play with numbers and experiment.

Multiplication by 5

As we have seen, to multiply by 5 we can multiply by 10 and then halve the answer. Five is half of 10. To multiply 6 by 5, we can multiply 6 by 10, which is 60, and then halve the answer to get 30.

Test yourself

Try these:

(a) 8 × 5 =

(b) 4 × 5 =

(c) 2 × 5 =

(d) 6 × 5 =

The answers are:

(a) 40   (b) 20   (c) 10   (d) 30

This is what we do when the tens digit is odd. Let's try 7 × 5. First multiply by 10:

$7 \times 10 = 70$

If you find it difficult to halve the 70, split it into 60 + 10. Half of 60 is 30 and half of 10 is 5, which gives us 35.

Let's try another:

$9 \times 5 =$

Ten nines are 90. Ninety splits to 80 + 10. Half of 80 + 10 is 40 + 5, so our answer is 45.

 Test yourself

Try these for yourself:

(a) $3 \times 5 =$

(b) $5 \times 5 =$

(c) $9 \times 5 =$

(d) $7 \times 5 =$

The answers are:

(a) 15   (b) 25   (c) 45   (d) 35

This is an easy way to learn the 5 times multiplication table.

This works for numbers of any size multiplied by 5. Let's try 14 × 5:

$14 \times 10 = 140$
$140 \div 2 = 70$

Let's try 23 × 5:

$23 \times 10 = 230$
$230 = 220 + 10$

Half of 220 + 10 is 110 + 5

$110 + 5 = 115$

These will become lightning mental calculations after just a few minutes' practice.

Experimenting with reference numbers

We use the reference numbers 10 and 100 because they are so easy to apply. It is easy to multiply by 10 (add a zero to the end of the number) and by 100 (add two zeros to the end of the number.) It also makes sense because your first step gives you the beginning of the answer.

We have also used 20 and 50 as reference numbers. Then we used 5 as a reference number to multiply very low numbers. In fact, we can use any number as a reference. We used 5 as a reference number to multiply 6 times 4 because we saw that 10 wasn't suitable.

Let's try using 9 as our reference number.

Subtract crossways:

$6 - 5 = 1$

or

$4 - 3 = 1$

Multiply 1 by our reference number, 9:

$1 \times 9 = 9$

Now multiply the numbers in the circles, and add the result to 9:

$3 \times 5 = 15$
$15 + 9 = 24$

The full working looks like this:

$$
\begin{array}{cccccc}
\boxed{9} & 6 & \times & 4 & = & \not{7} \\
 & \boxed{-3} & & \boxed{-5} & & 9 \\
 & & & & & +\underline{15} \\
 & & & & & 24 \quad \textit{Answer}
\end{array}
$$

It certainly is not a practical way to multiply 6 times 4 but you see that we can make the formula work if we want to. Let's try some more.

Let's try 8 as a reference number.

$$
\begin{array}{cccccc}
\boxed{8} & 6 & \times & 4 & = & \not{2} \\
 & \boxed{-2} & & \boxed{-4} & & 16 \\
 & & & & & +\underline{8} \\
 & & & & & 24 \quad \textit{Answer}
\end{array}
$$

Let's try 7 as a reference number.

$$
\begin{array}{ccccc}
\text{⑦} & 6 & \times & 4 & = & \cancel{3} \\
 & \text{(−1)} & & \text{(−3)} & & 21 \\
 & & & & & +\ \underline{3} \\
 & & & & & 24 \quad \textit{Answer}
\end{array}
$$

Let's try 3 as a reference number.

$$
\begin{array}{ccccc}
 & \text{(+3)} & & \text{(+1)} & & \\
\text{③} & 6 & \times & 4 & = & \cancel{7} \\
 & & & & & 21 \\
 & & & & & +\ \underline{3} \\
 & & & & & 24 \quad \textit{Answer}
\end{array}
$$

Let's try 2 as a reference number.

$$
\begin{array}{ccccc}
 & \text{(+4)} & & \text{(+2)} & & \\
\text{②} & 6 & \times & 4 & = & \cancel{8} \\
 & & & & & 16 \\
 & & & & & +\ \underline{8} \\
 & & & & & 24 \quad \textit{Answer}
\end{array}
$$

Try using the reference numbers 9 and 11 to multiply 7 by 8.

Let's try 9 first.

$$
\begin{array}{ccccc}
\text{⑨} & 7 & \times & 8 & = \\
 & \text{(−2)} & & \text{(−1)} & \\
\end{array}
$$

The numbers are below 9 so we draw the circles below. How much below 9? Two and 1. Write 2 and 1 in the circles. Subtract crossways.

$7 - 1 = 6$

Multiply 6 by the reference number, 9.

$6 \times 9 = 54$

This is our subtotal.

Now multiply the numbers in the circles and add the result to 54.

$2 \times 1 = 2$
$54 + 2 = 56$ *Answer*

$$\begin{array}{ccccccc}
\textcircled{9} & & 7 & \times & 8 & = & 6 \\
& & \textcircled{-2} & & \textcircled{-1} & & 54 \\
& & & & & & +\ \underline{2} \\
& & & & & & 56 \quad \textit{Answer}
\end{array}$$

This is obviously impractical but it proves it can be done.

Let's try 11 as a reference number.

$$\begin{array}{cccccc}
\textcircled{11} & & 7 & \times & 8 & = \\
& & \textcircled{-4} & & \textcircled{-3} &
\end{array}$$

Seven is 4 below 11 and 8 is 3 below. We write 4 and 3 in the circles.

Subtract crossways.

$7 - 3 = 4$

Multiply by the reference number.

$11 \times 4 = 44$

Multiply the numbers in the circles and add the result to 44.

$4 \times 3 = 12$

$44 + 12 = 56$ *Answer*

⑪   7   ×   8   =   44

(-4)   (-3)   + 12

56   *Answer*

This method will also work if we use 8 or 17 as reference numbers. It obviously makes more sense to use 10 as a reference number for the calculation as we can immediately call out the answer, 56. Still, it is always fun to play and experiment with the methods.

Try multiplying 8 times 8 and 9 times 9 using 11 as a reference number. This is only for fun. It still makes sense to use 10 as a reference number. Play and experiment with the method by yourself.

Chapter 8
Multiplication by 11

Everyone knows it is easy to multiply by 11. If you want to multiply 4 by 11, you simply repeat the digit for the answer, 44. If you multiply 6 by 11 you get 66. What answer do you get if you multiply 9 by 11? Easy, 99.

Did you know it is just as easy to multiply two-digit numbers by 11?

Multiplying a two-digit number by 11

To multiply 14 by 11 you simply add the digits of the number, $1 + 4 = 5$, and place the answer in between the digits for the answer, in this case giving you 154.

Let's try another: 23×11.

$2 + 3 = 5$

Answer: 253

These calculations are easy to do in your head. If somebody asks you to multiply 72 by 11 you could immediately say, 'Seven hundred and ninety-two.'

Test yourself

Here are some to try for yourself. Multiply the following numbers by 11 in your head:

42, 53, 21, 25, 36, 62

The answers are: 462, 583, 231, 275, 396 and 682.

They were easy, weren't they?

What happens if the digits add to 10 or more? It is not difficult — you simply carry the 1 (tens digit) to the first digit. For example, if you were multiplying 84 by 11 you would add 8 + 4 = 12, put the 2 between the digits and add the 1 to the 8 to get 924.

$$\overset{9}{84} \times 11 = \overset{}{\cancel{8}24}$$

If you were asked to multiply 84 by 11 in your head, you would see that 8 plus 4 is more than 10, so you could add the 1 you are carrying first and say, 'Nine hundred ...', before you even add 8 plus 4. When you add 8 and 4 to get 12, you take the 2 for the middle digit, followed by the remaining 4, so you would say '... and twenty- ... four.'

Let's try another. To multiply 28 by 11 you would add 2 and 8 to get 10. You add the carried 1 to the 2 and say, 'Three hundred and ...'. Two plus 8 is 10, so the middle digit is 0. The final digit remains 8, so you say, 'Three hundred and ... eight.'

How would we multiply 95 by 11?

Nine plus 5 is 14. We add 1 to the 9 to get 10. We work with 10 the same as we would with a single-digit number. Ten is the first part of our answer, 4 is the middle digit and 5 remains the final digit. We say, 'One thousand and ... forty-five.' We could also just say 'Ten ... forty-five.'

Practise some problems for yourself and see how quickly you can call out the answer. You will be surprised how fast you can call them out.

This approach also applies to numbers such as 22 and 33. Twenty-two is 2 times 11. To multiply 17 by 22, you would multiply 17 by 2 to get 34, then by 11 using the short cut, to get 374. Easy.

Multiplying larger numbers by 11

There is a very simple way to multiply any number by 11. Let's take the example of 123 × 11.

We would write the problem like this:

$\underline{0123} \times 11$

We write a zero in front of the number we are multiplying. You will see why in a moment. Beginning with the units digit, add each digit to the digit on its right. In this case, add 3 to the digit on its right. There is no digit on its right, so add nothing:

$3 + 0 = 3$

Write 3 as the last digit of your answer. Your calculation should look like this:

$$\frac{0123}{3} \times 11$$

Now go to the 2. Three is the digit on the right of the 2:

$$2 + 3 = 5$$

Write 5 as the next digit of your answer. Your calculation should now look like this:

$$\frac{0123}{53} \times 11$$

Continue the same way:

$$1 + 2 = 3$$
$$0 + 1 = 1$$

Here is the finished calculation:

$$\frac{0123}{1353} \times 11$$

If we hadn't written the 0 in front of the number to be multiplied, we might have forgotten the final step.

This is an easy method for multiplying by 11. The strategy develops your addition skills while you are using the method as a short cut.

Let's try another problem. This time we will have to carry digits. The only digit you can carry using this method is 1.

Let's try this example:

$$471 \times 11$$

We write the problem like this:

$\underline{0471} \times 11$

We add the units digit to the digit on its right. There is no digit to the right, so 1 plus nothing is 1. Write down 1 below the 1. Now add the 7 to the 1 on its right:

$7 + 1 = 8$

Write the 8 as the next digit of the answer. Your working so far should look like this:

$$\frac{0471}{81} \times 11$$

The next steps are:

$4 + 7 = 11$

Write 1 and carry 1. Add the next digits:

$0 + 4 + 1 (\text{carried}) = 5$

Write the 5, and the calculation is complete. Here is the finished calculation:

$$\frac{0471}{5^{1}181} \times 11$$

That was easy. The highest digit we can carry using this method is 1. Using the standard method to multiply by 11 you can carry any number up to 9. This method is easy and fun.

A simple check

Here is a simple check for multiplication by 11 problems. The problem isn't completed until we have checked it.

Let's check our first problem:

$$\frac{0123}{1353} \times 11$$

Write an X under every second digit of the answer, beginning from the right-hand end of the number. The calculation will now look like this:

$$\frac{0123}{1353} \times 11$$

$$X\,X$$

Add the digits with the X under them:

$$1 + 5 = 6$$

Add the digits without the X:

$$3 + 3 = 6$$

If the answer is correct, the answers will be the same, have a difference of 11, or a difference of a multiple of 11, such as 22, 33, 44 or 55. Both added to 6, so our answer is correct. This is also a test to see if a number can be evenly divided by 11.

Let's check the second problem:

$$\frac{0471}{5^{1}181} \times 11$$

$$X\ \ X$$

Add the digits with the X under them:

$$5 + 8 = 13$$

Add the digits without the X:

$$1 + 1 = 2$$

To find the difference between 13 and 2, we take the smaller number from the larger number:

13 − 2 = 11

If the difference is 0, 11, 22, 33, 44, 55, 66, etc., then the answer is correct. We have a difference of 11, so our answer is correct.

Note to parents and teachers

I often ask students in my classes to make up their own numbers to multiply by 11 and see how big a difference they can get. The larger the number they are multiplying, the greater the difference is likely to be. Let your students try for a new record.

Children will multiply a 700-digit number by 11 in their attempt to set a new record. While they are trying for a new world record, they are improving their basic addition skills and checking their work as they go.

Multiplying by multiples of 11

What are factors? We read about factors in Chapter 6. We found that it is easy to multiply by 20 because 2 and 10 are factors of 20, and it is easy to multiply by both 2 and 10.

It is easy to multiply by 22 because 22 is 2 times 11, and it is easy to multiply by both 2 and 11.

It is easy to multiply by 33 because 33 is 3 times 11, and it is easy to multiply by both 3 and 11.

Whenever you have to multiply any number by a multiple of 11 — such as 22 (2 × 11), 33 (3 × 11) or 44 (4 × 11) — you can combine the use of factors and the short cut for multiplication by 11. For instance, if you want to multiply 16 by 22, it is easy if you see 22 as 2 × 11. Then you multiply 16 by 2 to get 32, and use the short cut to multiply 32 by 11.

Let's try multiplying 7 by 33.

$7 \times 3 = 21$
$21 \times 11 = \mathbf{231}$ *Answer*

How about multiplying 14 by 33? Thirty-three is 3 times 11. How do you multiply 3 by 14? Fourteen has factors of 2 and 7. To multiply 3 × 14 you can multiply 3 × 2 × 7.

$3 \times 2 = 6$
$6 \times 7 = 42$
$42 \times 11 = \mathbf{462}$ *Answer*

You can think of the problem 14 × 33 as 7 × 2 × 3 × 11. Seven times 3 × 2 is the same as 7 × 6, which is 42. You are then left with 42 × 11. Teaching yourself to look for factors can make life easier and can turn you into a real mathematician.

One last question for this chapter: how would you multiply by 55? To multiply by 55 you could multiply by 11 and 5 (5 × 11 = 55) or you could multiply by 11 and 10 and halve the answer. Always remember to look for the method that is easiest for you.

After you have practised these strategies, you will begin to recognise the opportunities to use them for yourself.

Chapter 9
Multiplying decimals

What are decimals?

All numbers are made up of digits: 0, 1, 2, 3, 4, 5, 6, 7, 8 and 9. Digits are like letters in a word. A word is made up of letters. Numbers are made up of digits: 23 is a two-digit number, made from the digits 2 and 3; 627 is a three-digit number made from the digits 6, 2 and 7. The position of the digit in the number tells us its value. For instance, the 2 in the number 23 has a value of 2 tens, and the 3 has a value of 3 ones. Numbers in the hundreds are three-digit numbers: 435, for example. The 4 is the hundreds digit and tells us there are 4 hundreds (400). The tens digit is 3 and signifies 3 tens (30). The units digit is 5 and signifies 5 ones, or simply 5.

When we write a number, the position of each digit is important. The position of a digit gives that digit its place value.

When we write prices, or numbers representing money, we use a decimal point to separate the dollars from the cents. For example, $2.50 represents 2 dollars and 50 hundredths of a dollar. The first digit after the decimal represents tenths of a dollar. (Ten 10¢ coins make a dollar.) The second digit after the decimal represents hundredths of a dollar. (One hundred cents make a dollar.) So $2.50, or two and a half dollars, is the same as 250¢. If we wanted to multiply $2.50 by 4 we could simply multiply the 250¢ by 4 to get 1,000¢. One thousand cents is the same as $10.00.

Digits after a decimal point have place values as well. The number 3.14567 signifies three ones, then after the decimal point we have one tenth, four hundredths, five thousandths, six ten-thousandths, and so on. So $2.75 equals two dollars, seven tenths of a dollar and five hundredths of a dollar.

To multiply a decimal number by 10 we simply move the decimal point one place to the right. To multiply 1.2 by 10 we move the decimal one place to the right, giving an answer of 12. To multiply by 100, we move the decimal two places to the right. If there aren't two digits, we supply them as needed by adding zeros. So, to multiply 1.2 by 100, we move the decimal two places, giving an answer of 120.

To divide by 10, we move the decimal one place to the left. To divide by 100, we move the decimal two places to the left. To divide 14 by 100 we place the decimal after the 14 and move it two places to the left. The answer is 0.14. (We normally write a 0 before the decimal if there are no other digits.)

Now, let's look at general multiplication of decimals.

Multiplication of decimals

Multiplying decimal numbers is no more complicated than multiplying any other numbers. Let us take an example of 1.2 × 1.4.

We write down the problem as it is, but when we are working it out we ignore the decimal points.

$$\textcircled{10} \qquad \overset{\textcircled{+2}}{1.2} \quad \times \quad \overset{\textcircled{+4}}{1.4} \quad =$$

Although we write 1.2 × 1.4, we treat the problem as:

12×14 =

We ignore the decimal point in the calculation; we calculate 12 plus 4 is 16, times 10 is 160. Four times 2 is 8, plus 160 is 168.

The problem will look like this:

$$\textcircled{10} \qquad \overset{\textcircled{+2}}{1.2} \quad \times \quad \overset{\textcircled{+4}}{1.4} \quad = \quad 160$$
$$+ \quad \underline{\quad 8}$$
$$168 \quad \textit{Answer}$$

Our problem was 1.2 × 1.4, but we have calculated 12 × 14, so our work isn't finished yet. We have to place a decimal point in the answer. To find where we put the decimal, we look at the problem and count the number of digits after the decimals in the multiplication. There are two digits after the decimals: the 2 in 1.2 and the 4 in 1.4. Because there are two digits after the decimals in the problem, there must be two digits after the decimal in the answer. We count two places from the right and put the decimal between the 1 and the 6, leaving two digits after it.

1.68 *Answer*

An easy way to double-check this answer would be to approximate. That means, instead of using the numbers we were given, 1.2 × 1.4, we round off to 1.0 and 1.5. This gives us 1.0 times 1.5, which is 1.5, so we know the answer should be somewhere between 1 and 2. This tells us our decimal is in the right place. This is a good double-check. You should always make this check when you are multiplying or dividing using decimals. The check is simply: does the answer make sense?

Let's try another.

9.6 × 97 =

We write the problem down as it is, but work it out as if the numbers are 96 and 97.

$$96 - 3 = 93$$
$$93 \times 100 \, (\text{reference number}) = 9,300$$
$$4 \times 3 = 12$$
$$9,300 + 12 = 9,312$$

Where do we put the decimal? How many digits follow the decimal in the problem? One. That's how many digits should follow the decimal in the answer.

931.2 *Answer*

To place the decimal, we count the total number of digits following the decimals in both numbers we are multiplying. We will have the same number of digits following the decimal in the answer.

We can double-check the answer by estimating 10 times 90; from this we know the answer is going to be somewhere near 900, not 9,000 or 90.

If the problem had been 9.6 × 9.7, then the answer would have been 93.12. Knowing this can enable us to take some short cuts that might not be apparent otherwise. We will look at some of these possibilities shortly. In the meantime, try these problems.

Test yourself

Try these:

(a)   1.2 × 1.2 =   (b)   1.4 × 1.4 =

(c)   12 × 0.14 =   (d)   96 × 0.97 =

(e)   0.96 × 9.6 =   (f)   5 × 1.5 =

How did you go? The answers are:

(a) 1.44   (b) 1.96   (c) 1.68
(d) 93.12   (e) 9.216   (f) 7.5

What if we had to multiply 0.13 × 0.14?

$13 \times 14 = 182$

Where do we put the decimal? How many digits come after the decimal point in the problem? Four, the 1 and 3 in the first number and the 1 and 4 in the second. So we count back four digits in the answer. But, wait a minute; there are only three digits in the answer. What do we do? We have to supply the fourth digit. So, we count back three digits, then supply a fourth digit by putting a 0 in front of the number. Then we put the decimal point before the 0, so that we have four digits after the decimal.

The answer looks like this:

.0182

We can also write another 0 before the decimal, because there should always be at least one digit before the decimal. Our answer would now look like this:

0.0182

Let's try some more:

$0.014 \times 1.4 =$

$14 \times 14 = 196$

Where do we put the decimal? There are four digits after the decimal in the problem; 0, 1 and 4 in the first number and 4 in the second, so we must have four digits after the decimal in the answer. Because there are only three digits in our answer, we supply a 0 to make the fourth digit.

Our answer is 0.0196.

Test yourself

Try these for yourself:

(a)   22 × 2.4 =   (b)   0.48 × 4.8 =

(c)   0.048 × 0.48 =   (d)   0.0023 × 0.23 =

Easy, wasn't it? Here are the answers:

(a) 52.8   (b) 2.304   (c) 0.02304   (d) 0.000529

Beating the system

Understanding this simple principle can help us solve some problems that appear difficult using our method but can be adapted to make them easy. Here is an example.

$8 \times 79 =$

What reference number would we use for this one? We could use 10 for the reference number for 8, but 79 is closer to 100. Maybe we could use 50. The speeds maths method is easier to use when the numbers are close together. So, how do we solve the problem? Why not call the 8, 8.0?

There is no difference between 8 and 8.0. The first number equals 8; the second number equals 8 too, but it is accurate to one decimal place. The value doesn't change.

We can use 8.0 and work out the problem as if it were 80, as we did above. We can now use a reference number of 100. Let's see what happens:

(100)   8.0   ×   79   =

(-20)   (-21)

Now the problem is easy. Subtract diagonally.

79 − 20 = 59

Multiply 59 by the reference number (100) to get 5,900.

Multiply the numbers in the circles.

20 × 21 = 420

(To multiply by 20 we can multiply by 2 and then by 10.) Add the result to the subtotal.

5,900 + 420 = 6,320

The completed problem would look like this:

(100)   8.0   ×   79   =   5,900

(-20)   (-21)   +   420

6,320

Now, we have to place the decimal. How many digits are there after the decimal in the problem? One, the 0 we provided. So we count one digit back in the answer.

632.0 *Answer*

We would normally write the answer as 632.

Let's check this answer using estimation. Eight is close to 10 so we can round upwards.

$10 \times 79 = 790$

The answer should be less than, but close to, 790. It certainly won't be around 7,900 or 79. Our answer of 632 fits so we can assume it is correct.

We can double-check by casting out nines.

$$8 \quad \times \quad 79 \quad = \quad 632$$
$$8 \qquad 7 \qquad 2$$

Eight times 7 equals 56, which reduces to 11, then 2. Our answer is correct.

Let's try another.

$98 \times 968 =$

We write 98 as 98.0 and treat it as 980 during the calculation. Our problem now becomes 980 × 968.

$$\underset{(-20)}{\overset{(1,000)}{980}} \quad \times \quad \underset{(-32)}{968} \quad =$$

Our next step is:

$968 - 20 = 948$

Multiply by the reference number:

$948 \times 1,000 = 948,000$

Now multiply 32 by 20. To multiply by 20 we multiply by 2 and by 10.

$32 \times 2 = 64$
$64 \times 10 = 640$

It is easy to add 948,000 and 640.

$948,000 + 640 = 948,640$

We now have to adjust our answer because we were multiplying by 98, not 980.

Placing one digit after the decimal (or dividing our answer by 10), we get 94,864.0, which we simply write as 94,864.

Our full working would look like this:

$$
\begin{array}{cccc}
\boxed{1,000} & 980 & \times & 968 & = 948,000 \\
\boxed{-20} & & \boxed{-32} & + & \underline{640} \\
& & & & 948,640 \quad \textit{Answer}
\end{array}
$$

Let's cast out nines to check the answer.

$$
\begin{array}{ccccc}
\cancel{98.0} & \times & \cancel{968} & = & \cancel{94,864} \\
8 & & 14 & & 22 \\
& & 5 & & 4
\end{array}
$$

Five times 8 is 40, which reduces to 4. Our answer, 94,864, also reduces to 4, so our answer seems to be correct.

We have checked our answer, but casting out nines does not tell us if the decimal is in the correct position, so let's double-check by estimating the answer.

Ninety-eight is almost 100, so we can see if we have placed the decimal point correctly by multiplying 968 by 100 to get 96,800. This has the same number of digits as our answer so we can assume the answer is correct.

Test yourself

Try these problems for yourself:

(a) 9 × 82 =

(b) 9 × 79 =

(c) 9 × 77 =

(d) 8 × 75 =

(e) 7 × 89 =

The answers are:

(a) 738   (b) 711   (c) 693
(d) 600   (e) 623

That was easy, wasn't it? If you use your imagination you can use these strategies to solve any multiplication problem. Try some yourself.

Chapter 10

Multiplication using two reference numbers

The general rule for using a reference number for multiplication is that you choose a reference number that is close to both numbers being multiplied. If possible, you try to keep both numbers either above or below the reference number so you end up with an addition.

What do you do if the numbers aren't close together? What do you do if it is impossible to choose a reference number that is anywhere close to both numbers?

Here is an example of how our method works using two reference numbers:

$8 \times 37 =$

Firstly, we choose two reference numbers. The first reference number should be an easy number to use as a multiplier, such as 10 or 100. In this case we choose 10 as our reference number for 8.

The second reference number should be a multiple of the first reference number. That is, it should be double the first reference number, or three times, four times or even sixteen times the first reference number. In this case I would choose 40 as the second reference number as it equals 4 times 10.

We then write the reference numbers in brackets to the side of the problem, with the easy multiplier written first and the second number written as a multiple of the first. So, we would write 10 and 40 as (10×4), the 10 being the main reference number and the 4 being a multiple of the main reference.

The problem is written like this:

$(10 \times 4) \quad 8 \times 37 =$

Both numbers are below their reference numbers, so we draw the circles below.

$(10 \times 4) \qquad 8 \quad \times \quad 37 \quad =$

What do we write in the circles? How much below the reference numbers are the numbers we are multiplying? Two and 3, so we write 2 and 3 in the circles.

$(10 \times 4) \qquad 8 \quad \times \quad 37 \quad =$

$(-2) \qquad (-3)$

Here is the difference. We now multiply the 2 below the 8 by the multiplication factor, 4, in the brackets. Two times

4 is 8. Draw another circle below the 8 (under the 2) and write –8 in it.

The calculation now looks like this:

(10 × 4)　　　8　　×　　37　　=

(–2)　　　(–3)

(–8)

Now subtract 8 from 37.

37 – 8 = 29

Write 29 after the equals sign.

(10 × 4)　　　8　　×　　37　　=　　29

(–2)　　　(–3)

(–8)

Now we multiply 29 by the main reference number, 10, to get 290. Then we multiply the numbers in the top two circles (2 × 3) and add the result to 290.

(10 × 4)　　　8　　×　　37　　=　　290

(–2)　　　(–3)　　　+　　6

(–8)　　　　　296　　*Answer*

Let's try another one:

96 × 289 =

We choose 100 and 300 as our reference numbers. We set out the problem like this:

(100 × 3)96 × 289 =

What do we write in the circles? Four and 11. Ninety-six is 4 below 100 and 289 is 11 below 300.

(100 × 3)　　96　×　289　=

circles: (−4)　(−11)

Now multiply the 4 below 96 by the multiplication factor, 3. This gives us: 4 × 3 = 12. Draw another circle below 96 and write 12 in this circle.

The calculation now looks like this:

(100 × 3)　　96　×　289　=

circles: (−4)　(−11)

(−12)

Subtract 12 from 289.

$289 - 12 = 277$

Multiply 277 by the main reference number, 100.

$277 \times 100 = 27,700$

Now multiply the numbers in the original circles (4 and 11).

$4 \times 11 = 44$

Add 44 to 27,700 to get 27,744. This can easily be calculated in your head with just a little practice.

The full calculation looks like this:

(100 × 3)　　96　×　289　=　27,700

circles: (−4)　(−11)　　+　　44

(−12)　　　　27,744　*Answer*

The calculation part of the problem is easy. The only difficulty you may have is remembering what you have to do next.

If the numbers are above the reference numbers, we do the calculation as follows. We will take 12 × 124 as an example:

$$(10 \times 12)\ 12 \times 124 =$$

We choose 10 and 120 as reference numbers. Because the numbers we are multiplying are both above the reference numbers we draw the circles above. How much above? Two and 4, so we write 2 and 4 in the circles.

$$(10 \times 12) \quad 12 \quad \times \quad 124 \quad =$$

Now we multiply the number above the 12 (2) by the multiplication factor, 12.

$$2 \times 12 = 24$$

Draw another circle above the 12 and write 24 in it.

$$(10 \times 12) \quad 12 \quad \times \quad 124 \quad =$$

Now we add crossways.

$$124 + 24 = 148$$

Write 148 after the equals sign and multiply it by our main reference number, 10.

$148 \times 10 = 1{,}480$

(+24)

(+2)   (+4)

(10 × 12)   12   ×   124   =   1,480

Now multiply the numbers in the original circles, 2 × 4, and add the answer to the subtotal.

$2 \times 4 = 8$
$1{,}480 + 8 = 1{,}488$ *Answer*

The completed problem looks like this:

(+24)

(+2)   (+4)

(10 × 12)   12   ×   124   =   1,480
                                   +____8
                                    1,488   *Answer*

I try to choose reference numbers that keep both numbers either above or below the reference numbers, so that I have an addition rather than a subtraction at the end. For instance, if I had to multiply 9 times 83, I would choose 10 and 90 as reference numbers rather than 10 and 80. Although both would work, I want to keep the calculations as simple as possible.

Let's calculate the answer using both combinations, so you can see what happens.

Firstly, we will use 10 and 90.

(10 × 9)   9   ×   83   =   740

$\boxed{-1}$ $\boxed{-7}$ + _7_

$\boxed{-9}$ 747   *Answer*

That was straightforward. We multiplied 9 by 1 and wrote 9 in the bottom circle. Then we performed the following calculations:

$83 - 9 = 74$
$74 \times 10 = 740$
$1 \times 7 = 7$
$740 + 7 = 747$

Easy. Now let's use 10 and 80 as reference numbers.

                  $\boxed{+3}$

(10 × 8)   9   ×   83   =   750

$\boxed{-1}$ − _3_

$\boxed{-8}$ 747   *Answer*

There was not much difference in this case, but if you have a large number to subtract you may find it easier to choose your reference numbers so that you end your calculation with an addition.

You could also solve this problem using 10 and 100 as reference numbers. Let's try it.

(10 × 10)   9   ×   83   =   730

$\boxed{-1}$ $\boxed{-17}$ + _17_

$\boxed{-10}$ 747   *Answer*

There are many ways you can use these methods to solve problems. It is not a matter of which method is right but which method is easiest, depending on the particular problem you are trying to solve.

 Test yourself

Try these problems by yourself:

(a) 9 × 68 =

(b) 8 × 79 =

(c) 94 × 192 =

(d) 98 × 168 =

Write them down and calculate the answers, and then see if you can do them all in your head. With practice it becomes easy. Here are the answers:

(a) 612   (b) 632   (c) 18,048   (d) 16,464

Easy multiplication by 9

It is easy to use this method to multiply any number from 1 to 1,000 by 9.

Let's try an example:

$76 \times 9 =$

We use 10 as our reference for 9 and 80 as our reference for 76. We can rewrite the problem like this:

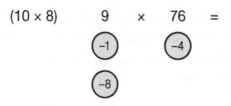

We subtract 8 from 76 (76 − 10 = 66, plus 2 is 68). Multiply 68 by the base reference, 10, which equals 680. Now multiply the numbers in the original circles, and add the result to 680.

$1 \times 4 = 4$

$680 + 4 = 684$ *Answer*

The base number is always 10 when you multiply by 9. Let's try another.

$9 \times 486 =$

The reference numbers are 10 and 490. You always go up to the next 10 for the second reference number. We write it like this:

(10 × 49)   9   ×   486   =
     ⊝−1   ⊝−4
     ⊝−49

We subtract 49 from 486. Is this difficult? Not at all. We subtract 50 and add 1:

$486 − 50 = 436$, plus 1 is 437

Multiply the numbers in the circles.

$1 \times 4 = 4$

This result is always the final digit in the answer when you multiply by 9. The answer is 4,374. The completed problem looks like this:

(10 × 49)   9   ×   486   =   4,370

       (−1)   (−4)   + ___4

      (−49)   4,374   *Answer*

The problems are easy to calculate in your head.

Using fractions as multiples

There is yet another possibility. The multiple can be expressed as a fraction.

What do I mean? Let's try 94 × 345. We can use 100 and 350 as reference numbers, which we would write as (100 × 3½). Our problem would look like this:

(100 × 3½)   94   ×   345   =

        (−6)   (−5)

Here we have to multiply −6 by 3½. To do this, simply multiply 6 by 3, which is 18, then add half of 6, which is 3, to get 21. Or, you could use factors and multiply 3½ by 2, and the resulting answer by 3 to get your answer of 21 (3½ × 2 is 7; 7 × 3 is 21).

(100 × 3½)   94   ×   345   =

        (−6)   (−5)

        (−21)

Subtract 21 from 345. Multiply the result by 100, then multiply the numbers in the circles, and then add these two results for our answer.

$345 - 21 = 324$
$324 \times 100 = 32,400$
$6 \times 5 = 30$
$32,400 + 30 = 32,430$ *Answer*

The completed problem looks like this:

$(100 \times 3\frac{1}{2})$ 94   ×   345   = 32,400
$\phantom{(100 \times 3\frac{1}{2})}$ (-6)   (-5)   +____30
$\phantom{(100 \times 3\frac{1}{2})}$ (-21)   32,430   *Answer*

Using factors expressed as division

To multiply 48 × 96, we could use reference numbers of 50 and 100, expressed as (50 × 2) or (100 ÷ 2). It would be easier to use (100 ÷ 2) because 100 then becomes the main reference number. It is easier to multiply by 100 than it is by 50.

When writing the multiplication, write the number first which has the main reference number, so instead of writing 48 × 96 we would write 96 × 48. The completed problem would look like this:

$(100 \div 2)$ 96   ×   48   = 4,600
 (-4)   (-2)   +____8
 (-2)   4,608   *Answer*

That worked out well, but what would happen if we multiplied 97 by 48? Then we have to halve an odd number. Let's see:

$$(100 \div 2) \quad 97 \quad \times \quad 48 \quad = \quad 4,650$$

$$\text{\small(-3)} \qquad \text{\small(-2)} \qquad + \quad \underline{6}$$

$$\text{\small(-}^3\!/_2\text{)} \qquad\qquad\qquad 4,656 \quad \textit{Answer}$$

$$\text{\small(-1}^1\!/_2\text{)}$$

In this case we have to divide the 3 below the 97 by the 2 in the brackets. Three divided by 2 is $^3/_2$ or $1\frac{1}{2}$. Subtracting $1\frac{1}{2}$ from 48 gives an answer of $46\frac{1}{2}$. Then we have to multiply $46\frac{1}{2}$ by 100. Forty-six by 100 is 4,600, plus half of 100 gives us another 50. So, $46\frac{1}{2}$ times 100 is 4,650.

Then we multiply 3 by 2 for an answer of 6; add this to 4,650 for our answer of 4,656.

Let us check this answer by casting out the nines:

```
 9̶7  ×  48  =  4̶,6̶5̶6
 7      12        12
         3         3
```

We check by multiplying 7 by 3 to get 21, which reduces to 3, the same as our answer. Our answer is correct.

What if we multiply 97 by 23? We could use 100 and 25 $(100 \div 4)$ as reference numbers.

$$(100 \div 4) \quad 97 \quad \times \quad 23 \quad =$$

$$\text{\small(-3)} \qquad \text{\small(-2)}$$

We divide 3 by 4 for an answer of ¾. Subtract ¾ from 23 (subtract 1 and give back ¼). We then multiply by 100.

$23 - \frac{3}{4} = 22\frac{1}{4}$

$22\frac{1}{4} \times 100 = 2,225 \left(25 \text{ is a quarter of } 100\right)$

The completed problem looks like this:

$(100 \div 4)$ 97   ×   23   =   2,225

           (−3)   (−2)   +   6

           (−³/₄)   2,231   *Answer*

You can use any combination of reference numbers. The general rules are:

1   Make the main reference number an easy number to multiply by; for example, 10, 20, 50 or 100.

2   The second reference number must be a multiple of the main reference number; for example, double the main reference number, or three times, ten times or fourteen times the main reference number.

There is no end to the possibilities. Play and experiment with the methods and you will find you are performing like a genius. Each time you use these strategies you develop your mathematical skills.

Would we use two reference numbers to calculate problems such as 8 × 17 or 7 × 26? No, I think the easy way to calculate 8 × 17 is to multiply 8 by 10 and then 8 by 7.

$8 \times 10 = 80$

$8 \times 7 = 56$

$80 + 56 = 136$

How about 7 × 26? I would say 20 times 7 is 140. To multiply by 20 we multiply by 2 and by 10. Two times 7 is 14, times 10 is 140. Then 7 times 6 is 42. The answer is 140 plus 42, which is 182.

Experiment for yourself to see which method you find easiest. You will find you can easily do the calculations entirely in your head.

Playing with two reference numbers

Let's multiply 13 times 76. We would use reference numbers of 10 and 70. We wouldn't normally use 80 as a reference number because we would have a subtraction at the end. Let's try a few methods to see how they work.

Firstly, using 10 and 70.

$$(10 \times 7) \qquad \overset{\textstyle \bigcirc\kern-1.1em {\scriptstyle +3}}{13} \quad \times \quad \overset{\textstyle \bigcirc\kern-1.1em {\scriptstyle +6}}{76} \quad =$$

We multiply the 3 above the 13 by the multiplication factor of 7 in the brackets. Three times 7 is 21; write 21 in a circle above the 3.

Now we add crossways.

76 + 21 = 97

Multiply 97 by the base reference number of 10 to get 970. Now multiply 3 times 6 to get 18. Add this to 970 to get 988.

Here is the full calculation:

$$(+21)$$

$$(+3) \qquad (+6)$$

$$(10 \times 7) \qquad 13 \quad \times \quad 76 \quad = \quad 970$$

$$+ \underline{\quad 18\quad}$$

$$988 \quad \textit{Answer}$$

Now let's try using reference numbers of 10 and 80.

$$(+3)$$

$$(10 \times 8) \qquad 13 \quad \times \quad 76 \quad =$$

$$(-4)$$

Multiply 3 times 8 to get 24. Write 24 in a circle above the 3. Add 24 to 76 to get 100. Multiply your answer by the base reference number of 10 to get 1,000.

The problem now looks like this:

$$(+24)$$

$$(+3)$$

$$(10 \times 8) \qquad 13 \quad \times \quad 76 \quad = \quad 1,000$$

$$(-4)$$

Multiply 3 times minus 4 to get minus 12. Subtract 12 from 1,000 to get your answer, 988. (To subtract 12, subtract 10, then 2.)

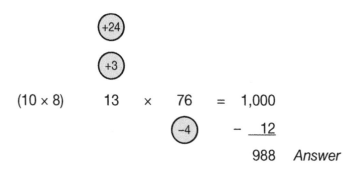

(10×8) 13 $\times$ 76 $=$ 1,000

$-$ <u>12</u>

988   *Answer*

Let's try another option; let's use 10 and 75 as reference numbers.

$(10 \times 7\frac{1}{2})$ 13 $\times$ 76 $=$

Thirteen is 3 above 10, and 76 is 1 above 75.

We multiply the 3 above the 13 by 7½. Is that difficult? No, we multiply 3 by 7 to get 21, plus half of 3 to get another 1½.

$21 + 1\frac{1}{2} = 22\frac{1}{2}$

Now we add crossways.

$76 + 22\frac{1}{2} = 98\frac{1}{2}$

Now we multiply 98½ by the base reference number of 10 to get 985. Then we multiply the numbers in the circles.

$3 \times 1 = 3$

Then we add 985 and 3 to get our answer.

$985 + 3 = 988$ *Answer*

The complete working looks like this:

$(+22½)$

$(+3)$ $(+1)$

$(10 × 7½)$ 13   ×   76   =   985
+ _3_
988   *Answer*

This is something to play with and experiment with. We just used the same formula three different ways to get the same answer. The last method (using a fraction as a multiplication factor) can be used to make many multiplication problems easier.

Let's try 96 times 321. We could use 100 and 325 as reference numbers.

$(100 × 3¼)$ 96   ×   321   =

(-4) (-4)

We multiply 4 by 3¼ to get 13 (4 × 3 = 12, plus a quarter of 4 gives another 1, making 13). Write 13 in a circle below the 4, under 96.

$(100 × 3¼)$ 96   ×   321   =

(-4) (-4)

(-13)

Subtract crossways.

321 − 13 = 308

Multiply 308 by the base reference number of 100 to get 30,800. Then multiply the numbers in the circles.

$4 \times 4 = 16$

Then:

$30,800 + 16 = 30,816$ *Answer*

That can easily be done in your head, and is most impressive.

Using decimal fractions as reference numbers

Here is another variation for using two reference numbers. The second reference number can be expressed as a decimal fraction of the first. Let's try it with 58 times 98. We will use reference numbers of 100 and 60, expressed as 0.6 of 100.

$(100 \times 0.6)98 \times 58 =$

The circles go below in each case. Write 2 inside both circles. Multiply 2 times the multiplication factor of 0.6. The answer is 1.2. All we do is multiply 6 times 2 to get 12, and divide by 10.

Our work looks like this:

(100×0.6) 98   ×   58   =

$\left(-2\right)$ $\left(-2\right)$

$\left(-1.2\right)$

Subtract 1.2 from 58 to get 56.8. Multiply by 100 to get 5,680. Multiply the numbers in the circles: 2 times 2 is 4.

$5,680 + 4 = 5,684$ *Answer*

Here is the problem fully worked out.

(100×0.6) 98   ×   58   =   5,680

(-2)   (-2)   +   4

(-1.2)   5,684   *Answer*

The above problem can be solved more easily by simply using a single reference number of 100 (try it), but it is useful to know there is another option.

The next example is definitely easier to solve using two reference numbers.

$96 \times 67 =$

We use reference numbers of 100 and 70.

(100×0.7) 96   ×   67   =

(-4)   (-3)

(-2.8)

We multiply 4 times 0.7 to get 2.8 ($4 \times 7 = 28$). We subtract 2.8 from 67 by subtracting 3 and adding 0.2 to get 64.2. Then we multiply our answer of 64.2 by 100 to get 6,420. Multiplying the circled numbers, 4 times 3, gives us 12.

$6,420 + 12 = 6,432$ *Answer*

That was much easier than using 100 as the reference number.

Here is another problem using 200 as our base reference number.

$189 \times 77 =$

We use 200 and 80 as our reference numbers. Eighty is 0.4 of 200.

$(200 \times 0.4) \quad 189 \quad \times \quad 77 \quad =$

(−11)　　　(−3)

(−4.4)

We multiply 11 by 0.4 to get 4.4. (We simply multiply 11 times 4 and divide the answer by 10.) We write 4.4 below the 11. We now subtract 4.4 from 77. To do this we subtract 5 and add 0.6. Seventy-seven minus 5 is 72, plus 0.6 is 72.6.

$72.6 \times 100 = 7,260$

We multiply 7,260 by 2 to get 14,520. (We could double 7 to get 14 and 2 times 26 is 52, making a mental calculation easy.) Our work so far looks like this:

$(200 \times 0.4) \quad 189 \quad \times \quad 77 \quad = \quad \cancel{7,260}$

(−11)　　　(−3)　　　　　　14,520

(−4.4)

Now we multiply the numbers in the circles.

$11 \times 3 = 33$

Add 33 to 14,520 to get 14,553. Here is the calculation worked out in full.

(200×0.4) 189   ×   77   =   ~~7,260~~

-11   -3   14,520

-4.4   +   33

14,553   *Answer*

It is interesting and fun to try different strategies to find the easiest method.

Test yourself

Try these for yourself:

(a)   92 × 147 =   (b)   88 × 172 =

(c)   94 × 68 =   (d)   96 × 372 =

The answers are:

(a) 13,524   (b) 15,136   (c) 6,392   (d) 35,712

How did you go? I used $100 \times 1\frac{1}{2}$ for (a), $100 \times 1\frac{3}{4}$ for (b), 100×0.7 for (c) and $100 \times 3\frac{3}{4}$ for (d).

By now you must find these calculations very easy. Experiment for yourself. Make up your own problems.

Try to solve them without writing anything down. Check your answers by casting out the nines.

What you are doing is truly amazing. You are certainly developing exceptional skills. Keep playing with the methods and you will find it easier with practice to do everything in your head. You are developing great mental and mathematical skills.

Chapter 11

Using factors with two reference numbers

One day I was being interviewed on a talk radio program, and I was talking about how students could master multiplication tables in minutes. I explained how I multiply 7 times 8 and 96 times 97 using circles.

One of the interviewers said, 'We want to know how you would multiply 26 times 37.' I said I would have to take them through a couple of intermediate explanations to explain it, but they insisted I explain how to multiply 26 times 37. I was not allowed to give my usual explanation; 'Just do the calculation,' they told me. I had several options, using 20 or 30 as a reference, or even using two reference numbers, 20 and 40.

My choice didn't go down well and my explanation fell flat — which, I suspect, was what they wanted to happen.

I have since realised how I should have answered. Almost any combination of numbers is easy to multiply if you use factors.

Multiplying using factors

Let's use the example I was given.

An easy way to multiply 26 times 37 is to break up 26 into factors of 2 × 13. Then our problem would look like this:

$13 \times 2 \times 37 \, or \, 13 \times (2 \times 37)$

If we multiply 2 × 37 we get 74. Our calculation is now 13 × 74. This is easily solved using 10 and 70 as reference numbers.

$$(10 \times 7) \quad \overset{\textstyle\bigcirc\kern-1.2em{+3}}{13} \quad \times \quad \overset{\textstyle\bigcirc\kern-1.2em{+4}}{74} \quad =$$

Now we multiply the 3 above 13 by the multiplication factor (in the brackets), 7, to get 21. We draw another circle above the 3 and write 21 in it.

$$(10 \times 7) \quad \overset{\textstyle\overset{(+21)}{(+3)}}{13} \quad \times \quad \overset{(+4)}{74} \quad =$$

We add 21 to 74 to get 95. Multiply 95 by the base reference number, 10, to get 950. This is our subtotal.

$95 \times 10 = 950$

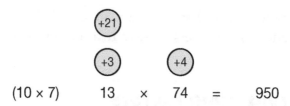

(10 × 7)   13 × 74 =   950

Now multiply the numbers in the original circles, 3 × 4, and add the answer to the subtotal.

3 × 4 = 12

950 + 12 = 962 *Answer*

The completed problem looks like this:

(+21)

(+3)   (+4)

(10 × 7)   13 × 74 =   950

+ 12

962   *Answer*

This calculation can easily be done in your head.

If you know the 13 times table you can calculate the answer even more quickly and with less effort. Thirteen times 7 is 91, so 13 times 70 is 910. Thirteen times 4 is 52. It is a simple task to add 52 to 910 in your head. Nine hundred and ten plus 50 is 960, plus 2 is 962.

Let's try another example, 38 × 73.

We can halve 38 to get 19, so the factors are 2 and 19. That would give us 73 × 2 × 19. Multiplying 73 by 2 we get 146, so the problem becomes 19 × 146. We can use

20 as a reference number and 140 (20 × 7) as our second reference number.

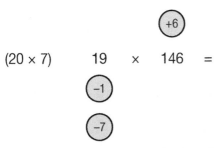

(20 × 7)   19   ×   146   =

We multiply 1 times 7 in the brackets to get the number in the second circle below 19. We subtract 7 from 146 to get 139.

To multiply 139 by 20, we double and then add a zero. (It is easy to work out what 139 doubled is — it is double 140 minus 1, which is 280 – 2 = 278.)

Multiply 278 by 10 to get 2,780.

Now multiply the numbers in the original circles:

$6 \times -1 = -6$

$2,780 - 6 = 2,774$ *Answer*

The completed problem looks like this:

(20 × 7)   19   ×   146   =   139

2,780

-   6

2,774   *Answer*

Another way to calculate this could be to halve 19 to get 9.5. The problem then becomes 9.5 × 292.

Now we can use 10 and 300 as reference numbers.

(10 × 30)   9.5   ×   292   =   2,770

(−0.5)   (−8)   +   __4__

(−15)   2,774   *Answer*

We perform the following calculations to get the answer:

$30 \times 0.5 = 15$
$292 - 15 = 277$
$277 \times 10 = 2,770$
$-0.5 \times -8 = 4$
$2,770 + 4 = 2,774$ *Answer*

Using factors gives us a greater choice of methods for using our formula. Both of the methods are easier than direct multiplication.

Using the factors method with prime numbers

Prime numbers are numbers that only have 1 and the number itself as potential factors; they have no other divisors. We just multiplied by 19, a prime number, by using factors 9.5 and 2. You can always divide a prime number by 2 to simplify the calculation.

Let's try 31 × 73, two prime numbers.

We could double 31 to get 62 and then halve the answer, but it would be easier to halve 31 to get 15.5 and multiply 73 by 2.

To multiply 73 by 2, you would break up 73 into 70 plus 3. If you see 70 as 7 times 10 you would multiply 7 by 2 to get 14 and then by 10 to get 140. Two times 3 is 6. Add six to 140 for the answer, 146.

The problem now becomes 15.5 by 146.

We use 10 and 140 as our reference numbers.

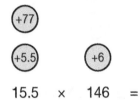

(10 × 14)   15.5   ×   146   =

Multiply 5.5 (the number above 15.5) by the multiplication factor, 14.

$5.5 \times (2 \times 7) =$
$5.5 \times 2 = 11$
$11 \times 7 = 77$

Write 77 in another circle above 15.5.

(10 × 14)   15.5   ×   146   =

Now we add crossways.

$77 + 146 = 223$

(Add 80 and then subtract 3, or add 100 and then subtract 20 and then subtract 3. With practice, you will find this easy and be able to do it easily in your head.)

Multiply by 10 to get a subtotal of 2,230.

Now multiply the numbers in the original circles.

$6 \times 5.5 = 33$
$2,230 + 33 = 2,263$ *Answer*

An alternative method for multiplying 31 by 73 would be to multiply 73 by 30 and then add 73 for the answer.

Let's try another example, 96 × 194.

To multiply 96 by 194, you can factorise 194 as 2 × 97.

We multiply 96 × 97 to get 9,312; 9,312 doubled is 18,624.

For an easier way to calculate this, use two reference numbers, 100 and 200.

The calculation looks like this:

(100×2) 96   ×   194   =

     (−4)   (−6)

     (−8)

Subtract 8 from 194 to get 186. Multiply 186 by the base reference number, 100, to get 18,600. Then multiply the numbers in the first two circles, −4 × −6 = 24.

$18,600 + 24 = 18,624$ *Answer*

The full calculation looks like this:

(100×2) 96   ×   194   =   18,600

     (−4)   (−6)   +   24

     (−8)   18,624   *Answer*

Test yourself

Try these problems for yourself:

(a) 24 × 37 =   (b) 71 × 26 =

(c) 33 × 72 =   (d) 28 × 54 =

Here are the answers:

(a) 888   (b) 1,846   (c) 2,376   (d) 1,512

With imagination and a little experimentation you can use this method to multiply any numbers.

Welcome challenges; every time someone challenged my methods it resulted in improvement.

Chapter 12
Addition

Most of us find addition easier than subtraction. We will learn in this chapter how to make addition even easier.

Some numbers are easy to add. It is easy to add numbers like 1 and 2, 10 and 20, 100 and 200, 1,000 and 2,000.

Twenty-five plus 1 is 26.

Twenty-five plus 10 is 35.

Twenty-five plus 100 is 125.

Twenty-five plus 200 is 225.

These are obviously easy, but how about adding 90? The easy way to add 90 is to add 100 and subtract 10.

Forty-six plus 90 equals 146 minus 10. What is 146 minus 10? The answer is 136.

$$46 + 90 = 146 - 10 = 136$$

What is the easy way to add 9? Add 10 and subtract 1.

What is the easy way to add 8? Add 10 and subtract 2.

What is the easy way to add 7? Add 10 and subtract 3.

What is the easy way to add 95? Add 100 and subtract 5.

What is the easy way to add 85? Add 100 and subtract 15.

What is the easy way to add 80? Add 100 and subtract 20.

What is the easy way to add 48? Add 50 and subtract 2.

Test yourself

Try the following to see how easy this is. Call out the answers as fast as you can.

24 + 10 =

38 + 10 =

83 + 10 =

67 + 10 =

It is easy to add 10 to any number. I'm sure I don't need to give you the answers to these.

Now, try the following problems in your head. Call out the answers as quickly as you can. Try to say the answer in one step. For example, for 34 + 9, you wouldn't call out, 'Forty-four … forty-three,' you would make the adjustment while you call out the answer. You would just say, 'Forty-three.' Try them. As long as you remember to add 9 by adding 10 and taking 1, add 8 by adding 10 and taking 2, and add 7 by adding 10 and taking 3, you will find them easy.

Test yourself

Try these:

(a)   25 + 9 =

(b)   46 + 9 =

(c)   72 + 9 =

(d)   56 + 8 =

(e)   37 + 8 =

(f)   65 + 7 =

The answers are:

(a) 34   (b) 55   (c) 81
(d) 64   (e) 45   (f) 72

You don't have to add and subtract the way you were taught at school. In fact, high maths achievers usually use different methods to everyone else. That is what makes them high achievers — not their superior brain.

How would you add 38? Add 40 and subtract 2.

So, how would you add the following?

   (a) 23 + 48 =
   (b) 126 + 39 =
   (c) 47 + 34 =
   (d) 424 + 28 =

For (a), you would say 23 plus 50 is 73, minus 2 is 71.

For (b), you would say 126 plus 40 is 166, minus 1 is 165.

For (c), you would say 50 plus 34 is 84, minus 3 is 81.

And for (d), you would say 424 plus 30 is 454, minus 2 is 452.

Did you find these easy to do in your head?

What if you have to add 31 to a number? You simply add 30, and then add the 1. To add 42 you add 40, and then add 2.

Test yourself

Try these:

(a)   26 + 21 =   (b)   43 + 32 =

(c)   64 + 12 =   (d)   56 + 41 =

The answers are:

(a) 47   (b) 75   (c) 76   (d) 97

You may have thought that all of those answers were obvious, but many people never try to calculate these types of problems mentally.

Often you can round numbers off to the next hundred. How would you do the following problem in your head?

2,351
+489

You could say 2,351 plus 500 is 2,851 (300 + 500 = 800), minus 11 is 2,840.

Normally you would use a pen and paper to make the calculation, but when you see 489 as 11 less than 500, the calculation becomes easy to do in your head.

Test yourself

Try these problems in your head:

(a)   531
     +297

(b)   333
     +249

(c) 4,537
     +388

For (a), you simply add 300 and subtract 3:

531 + 300 is 831, minus 3 is 828

For (b), you add 200, then add 50, and subtract 1:

333 + 200 is 533, plus 50 is 583, minus 1 is 582

For (c), you add 400 and subtract 12:

4,537 plus 400 is 4,937, minus 12 is 4,925

Adding from left to right

For most additions, if you are adding mentally, you should add from left to right instead of from right to left as you are taught in school.

How would you add these numbers in your head?

    5,164
    +2,938

I would add 3,000, then subtract 100 to add 2,900. Then I would add 40 and subtract 2 to add 38.

To add 2,900 you would say, 'Eight thousand and sixty-four,' then add 40 and subtract 2 to get 'Eight thousand, one hundred and two.'

This strategy makes it easy to keep track of your calculation and hold the numbers in your head.

Order of addition

Let's say we have to add the following numbers:

    6
    8
    +4

The easy way to add the numbers would be to add:

6 + 4 = 10, plus 8 = 18

Most people would find that easier than 6 + 8 + 4 = 18.

So, an easy rule is, when adding a column of numbers, add pairs of digits to make tens first if you can, and then add the other digits.

You can also add a digit to make up the next multiple of 10. That is, if you have reached, say, 27 in your addition, and the next two numbers to add are 8 and 3, add the 3 before the 8 to make 30, and then add 8 to make 38. Using our methods of multiplication will help you to remember the combinations of numbers that add to 10, and this should become automatic.

This also applies if you had to solve a problem such as 26 + 32 + 14. We can see that the units digits from 26 and 14 (6 and 4) add to 10, so it is easy to add 26 and 14. I would add 26 plus 10 to make 36, plus 4 makes 40, and then add the 32 for an answer of 72. This is an easy calculation compared with adding the numbers in the order they are written.

Breakdown of numbers

You also need to know how numbers are made up. Twelve is not only 10 plus 2 but also 8 plus 4, 6 plus 6, 7 plus 5, 9 plus 3 and 11 plus 1. It is important that you can break up all numbers from 1 to 10 into their basic parts.

$$2 = \quad 1+1$$
$$3 = \quad 2+1$$
$$4 = \quad 2+2, 3+1$$
$$5 = \quad 3+2, 4+1$$
$$6 = \quad 3+3, 4+2, 5+1$$
$$7 = \quad 4+3, 5+2, 6+1$$
$$8 = \quad 4+4, 5+3, 6+2, 7+1$$
$$9 = \quad 5+4, 6+3, 7+2, 8+1$$
$$10 = \quad 5+5, 6+4, 7+3, 8+2, 9+1$$

How do you add 8 plus 5? You could add 5 plus 10 and subtract 2. Or you could add 8 plus 2 to make 10, then add another 3 (to make up the 5) to give the same answer of 13.

How do you add 7 + 6? Six is 3 + 3. Seven plus 3 is 10, plus the second 3 is 13.

Note to parents and teachers

Often children are just told, 'You simply have to memorise the answers to 8 plus 5 or 8 plus 4.' And with practice they will be memorised, but give them a strategy to calculate the answer in the meantime.

Checking addition by casting out nines

Just as we cast out nines to check our answers for multiplication, so we can use the strategy for checking our addition and subtraction.

Here is an example:

12345
67890
41735
+21865

We add the numbers to get our total of 143,835.

Is our answer correct? Let's check by casting out nines, or working with our substitute numbers.

1234~~5~~	6	
67~~89~~0	21	3
~~4~~173~~5~~	11	2
+ 2~~1~~86 5	13	<u>4</u>
~~1~~438~~35~~	6	

Our substitutes are 6, 3, 2 and 4. The first 6 and 3 cancel (6 + 3 = 9) to leave us with just 2 and 4 to add.

2 + 4 = 6

Six is our check or substitute answer.

The real answer should add to 6. Let's see. After casting nines, we have:

~~1~~ + 4 + 3 + ~~8~~ + 3 + ~~5~~ = 6

Our answer is correct.

But there is a further short cut when you are casting nines to check a calculation in addition. You can cast out (cross out) any digits anywhere in the addition that add to 9. You can't, of course, cross out digits in the answer with digits you are adding. So, you could have cast out the 6 in the fourth number with the 3 in the third, the 7 in the third number with the 2 in the fourth. You could have cast out every digit in the top two numbers as each digit in the top number combines with a digit in the second to equal 9. Your check could look like this:

~~12345~~ 0
~~67890~~ 0
4 1~~7 3 5~~ 1
+ ~~2 1~~8 6 5 <u>5</u>
~~1~~43~~8~~3~~5~~ 6

Note that this only works with addition. With all other mathematical checks you must find substitutes for each number. With addition you can cross out (cast out) any digits of the numbers you are adding. But remember, you can't mix digits in the numbers you are adding with digits in the answer.

Let's look at another example.

234
671
+ <u>855</u>
1,760

Here the 2 in 234 can be crossed with the 7 in 671. The 3 in 234 combines with the 6 in 671. The 4 in 234 combines with a 5 in 855. The 1 in 671 combines with the 8 in 855. All digits are cast out except a 5 in 855. That means the answer must add to 5. Let's now check the answer:

$1 + 7 + 6 + 0 = 14$
$1 + 4 = 5$

Our calculation is correct. The final check looks like this:

```
  2̶3̶4̶   0
  6̶7̶1̶   0
+ 8̶5̶5    5
  1 7 6 0   5
```

If the above figures were amounts of money with a decimal, it would make no difference. You can use this method to check almost all of your additions, subtractions, multiplications and divisions. And you can have fun doing it.

Note to parents and teachers

The basic number facts are quickly learnt when children master these techniques and learn the multiplication tables.

Casting out nines to check answers will give plenty of practice with the combinations of numbers that add to 9. These exercises don't need to be drilled. They are learned naturally when students use these strategies.

Once a student knows the basic make-up of numbers, he or she can easily add small numbers without mistake and without borrowing.

Chapter 13
Subtraction

Most of us find addition easier than subtraction. Subtraction, the way most people are taught in school, is more difficult. It need not be so. You will learn some strategies in this chapter that will make subtraction easy.

Firstly, you need to know the combinations of numbers that add to 10. You learnt those when you learnt the speed maths method of multiplication. You don't have to think too hard; when you multiply by 8, what number goes in the circle below? You don't have to calculate; you don't have to subtract 8 from 10. You know a 2 goes in the circle from so much practice. It is automatic.

If a class is asked to subtract 9 from 56, some students will use an easy method and give an immediate answer. Because their method is easy, they will be fast and unlikely to make a mistake. The students who use a difficult method will take longer to solve the problem and, because their method is difficult, they are more likely to make a mistake. Remember my rule:

The easiest way to solve a problem is also the fastest, with the least chance of making a mistake.

So, what is the easy way to subtract 9? Subtract 10 and add (give back) 1.

What is the easy way to subtract 8? Subtract 10 and add (give back) 2.

What is the easy way to subtract 7? Subtract 10 and add (give back) 3.

What is the easy way to subtract 6? Subtract 10 and add (give back) 4.

What is the easy way to subtract 90? Subtract 100 and add (give back) 10.

What is the easy way to subtract 80? Subtract 100 and add (give back) 20.

What is the easy way to subtract 70? Subtract 100 and add (give back) 30.

What is the easy way to subtract 95? Subtract 100 and add (give back) 5.

What is the easy way to subtract 85? Subtract 100 and add (give back) 15.

What is the easy way to subtract 75? Subtract 100 and add (give back) 25.

What is the easy way to subtract 68? Subtract 70 and add (give back) 2.

Do you get the idea? The simple way to subtract is to round off and then adjust.

What is 284 minus 68? Let's subtract 70 from 284 and add 2.

$$284 - 70 = 214$$
$$214 + 2 = 216$$

This is easily done in your head. Calculating with a pencil and paper involves carrying and borrowing. This way is much easier.

How would you calculate 537 minus 298? Again, most people would use pen and paper. The easy way is to subtract 300 and give back 2.

$$537 - 300 = 237$$
$$237 + 2 = 239$$

Using a written calculation the way people are taught in school would mean carrying and borrowing twice.

To subtract 87 from a number, take 100 and add 13 (because 100 is 13 more than you wanted to subtract).

$$432 \quad - \quad 87 \quad =$$

$$\textcircled{13}$$

Subtract 100 to get 332. Add 13 (add 10 and then 3) to get 345. Easy.

Test yourself

Try these problems in your head. You can write down the answers.

(a) 86 − 38 =   (b) 42 − 9 =

(c) 184 − 57 =   (d) 423 − 70 =

(e) 651 − 185 =   (f) 3,424 − 1,895 =

The answers are:

(a) 48   (b) 33   (c) 127
(d) 353   (e) 466   (f) 1,529

For (a) you would subtract 40 and add 2.

For (b) you would subtract 10 and add 1.

For (c) you would subtract 60 and add 3.

For (d) you would subtract 100 and add 30.

For (e) you would subtract 200 and add 15.

For (f) you would subtract 2,000 and add 100, then 5.

In each case there is an easy subtraction — and the rest is addition.

Numbers around 100

When we subtract a number just below 100 from a number just above 100, there is an easy method. You can draw a circle below the number you are subtracting and write in the amount you need to make 100. Then add the number in the circle to the amount the first number is above 100. This turns subtraction into addition.

Let's try one.

$$\overset{\displaystyle(23)}{123} \quad - \quad \underset{\displaystyle(25)}{75} \quad =$$

Seventy-five is 25 below 100. Add 25 to 23 for an easy answer of 48. Let's try another.

$$\overset{\displaystyle(32)}{132} \quad - \quad \underset{\displaystyle(12)}{88} \quad =$$

$$12 + 32 = 44 \qquad \textit{Answer}$$

This technique works for any numbers above and below any hundreds value.

$$\overset{\displaystyle(64)}{364} \quad - \quad \underset{\displaystyle(22)}{278} \quad =$$

$$64 + 22 = 86 \qquad \textit{Answer}$$

It also works for subtracting numbers near the same tens value.

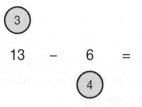

3 + 4 = 7   *Answer*

Here's another:

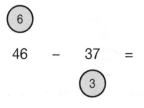

6 + 3 = 9   *Answer*

Test yourself

Try these for yourself:

(a) 13 − 8 =   (b) 17 − 8 =

(c) 12 − 9 =   (d) 12 − 8 =

(e) 124 − 88 =   (f) 161 − 75 =

(g) 222 − 170 =

(h) 111 − 80 =

(i) 132 − 85 =

(j) 145 − 58 =

How did you go? They are easy if you know how. Here are the answers:

(a) 5	(b) 9	(c) 3	(d) 4
(e) 36	(f) 86	(g) 52	(h) 31
(i) 47	(j) 87		

If you made any mistakes, go back and read the explanation and try them again.

Easy written subtraction

Easy subtraction uses either of two carrying and borrowing methods. You should recognise one or even both methods.

The difference between standard subtraction and easy subtraction is minor, but important. I will explain easy subtraction with two methods of carrying and borrowing. Use the method you are familiar with or that you find easier.

Subtraction method one

Here is a typical subtraction:

```
 8265
−3897
 4368
```

This is how the working might look:

```
  715
 8265
-3897
 4368
```

Let's see how easy subtraction works. Subtract 7 from 5. You can't, so you 'borrow' 1 from the tens column. Cross out the 6 and write 5. Now, here is the difference. You don't say 7 from 15, you say 7 from 10 equals 3, then add the number above (5) to get 8, the first digit of the answer.

With this method, you never subtract from any number higher than 10. The rest is addition.

Nine from 5 won't go, so borrow again. Nine from 10 is 1, plus 5 is 6, the next digit of the answer.

Eight from 1 won't go, so borrow again. Eight from 10 is 2, plus 1 is 3, the next digit of the answer.

Three from 7 is 4, the final digit of the answer.

Subtraction method two

```
  8,2,6,5
-,3,8,9 7
  4 3 6 8
```

Subtract 7 from 5. You can't, so you borrow 1 from the tens column. Put a 1 in front of the 5 to make 15 and write a small 1 alongside the 9 in the tens column. Using our easy method, you don't say 7 from 15, but 7 from 10 is 3, plus 5 on top gives 8, the first digit of the answer.

Ten (9 plus 1 carried) from 6 won't go, so we have to borrow; 10 from 10 is 0, plus 6 is 6.

Nine from 2 won't go, so we borrow again. Nine from 10 is 1, plus 2 is 3.

Four from 8 is 4. We have our answer.

You don't have to learn or know the combinations of single-digit numbers that add to more than 10. You never subtract from any number higher than 10. Most of the calculation is addition. This makes the calculations easier and reduces mistakes.

Test yourself

Try these for yourself:

(a)   7,235
       −4,568

(b)   5,417
       −3,179

The answers are:

(a) 2,757

(b) 2,238

Note to parents and teachers

This strategy is very important. If a student has mastered multiplication using the simple formula in this book, they have mastered the combinations of numbers that add to 10. There are only five such combinations.

(continued)

If a student has to learn the combinations of single-digit numbers that add to more than 10, there are another twenty such combinations to learn. Using this strategy, children don't need to learn any of them. To subtract 8 from 15, they can subtract from 10 (which gives 2), and then add the 5, for an answer of 7.

There is a far greater chance of making a mistake when subtracting from numbers in the teens than when subtracting from 10. There is very little chance of making a mistake when subtracting from 10; when children have been using the methods in this book the answers will be almost automatic.

Subtraction from a power of 10

There is an easy method for subtraction from a number ending in several zeros. This can be useful when using 100 or 1,000 as reference numbers. The rule is:

Subtract the units digit from 10, then each successive digit from 9, then subtract 1 from the digit to the left of the zeros.

For example:

```
 1000
−368
```

We can begin from the left or right.

Let's try it from the right first. Subtract the units digit from 10.

$$10 - 8 = 2$$

This is the right-hand digit of the answer. Then take the other digits from 9.

Six from 9 is 3.

Three from 9 is 6.

One from 1 is 0.

So we have our answer: 632.

Now let's try it from left to right.

One from 1 is 0. Three from 9 is 6. Six from 9 is 3. Eight from 10 is 2. Again we have 632.

Here is what we really did. The set problem was:

```
  1000
-  368
  632
```

We subtracted 1 from the number we were subtracting from, 1,000, to get 999. We then subtracted 368 from 999 with no numbers to carry and borrow (because no digits in the number we are subtracting can be higher than 9). We compensated by adding the 1 back to the answer by subtracting the final digit from 10 instead of 9.

So what we really calculated was this:

```
 999+1
-368
 631+1
```

This simple method makes a lot of subtraction problems much easier.

If you had to calculate 40,000 minus 3,594, this is how you would do it:

40000
−03594

Take 1 from the left-hand digit (4) to get 3, the first digit of the answer. (We are really subtracting 3,594 from 39,999 and then adding the extra 1.)

Three from 9 is 6. Five from 9 is 4. Nine from 9 is 0. Four from 10 is 6.

The answer is 36,406.

You could do the calculation off the top of your head. You would call out the answer, 'Thirty-six thousand, four hundred and six.' Try it. With a little practice you can call out the digits without a pause. You may find it easier just to say, 'Three, six, four, oh, six.' Either way, it is very impressive.

Test yourself

Try these for yourself:

(a)   10,000 − 2,345   (b)   60,000 − 41,726

You would have found the answers are:

(a) 7,655   (b) 18,274

Subtracting smaller numbers

If the number you are subtracting is short, then add zeroes before the number you are subtracting (at least mentally) to make the calculation.

Let's try 45,000 – 23:

```
  45000
 –0023
  44977
```

You extend the zeroes in front of the subtrahend (the number being subtracted) as far as the first digit that is not a 0 in the top number. You then subtract 1 from this digit. (Again, think of 45,000 as 44,999 plus 1.) Five minus 1 equals 4.

Subtract each successive digit from 9 until you reach the final digit, which you subtract from 10.

An easy way to understand the procedure is to subtract 1 from both numbers. The answer will be the same but there are no numbers to carry or borrow. Subtracting 1 (or any given number) from both numbers alters the calculation but the answer remains the same. Eighteen minus 13 gives an answer of 5. Subtracting 1 from both numbers (17 – 12) still gives an answer of 5. Subtracting 2 from each number gives the same answer: 16 – 11 = 5. You may prefer to mentally subtract 1 from the numbers to make the calculation in your head.

This strategy is useful for calculating the numbers to write in the circles when you are using 100 or 1,000 as reference numbers for multiplication. It is also useful for calculating change.

The method taught in North American and Australian schools has you doing exactly the same calculation, but you have to work out what you are carrying and borrowing with each step. The benefit of this method is that it becomes mechanical and can be carried out with less chance of making a mistake.

An alternate method of subtraction

We can use the above strategy of subtracting from numbers ending in zeroes as an alternate method for subtraction with any numbers.

Imagine you have $1.20 in your pocket and you spy your favourite chocolate bar on sale reduced to 80 cents. You take it to the checkout and put down a dollar and get 20 cents change. The 20 cents change plus the 20 cents in your pocket makes 40 cents. You have subtracted 80 from 120 by subtracting 80 from 100 and adding 20 to the answer.

Let's see how it works in practice.

Example:

$456 - 372 =$

Instead of subtracting 372 from 456, we subtract 372 from 400 and add the 56.

$$400$$
$$-\underline{372}$$

We call the 400 as 399 + 1 and subtract the 7 from 9 and the 2 from 10.

We get an answer of 28. It is easy to add 28 to 56 for the answer of 84. There was no carrying or borrowing.

Let's try it with larger numbers.

```
 4372
-2198
```

Subtract 2,198 from 3,999 + 1 and we get our working answer of 1,802.

```
   +1
 3999
-2198
 1802
```

It is easy to add 1,802 and 372 to get 2,174. (Add 1,800 plus 300 you get 2,100, then 72 plus 2 for the answer 2,174.)

```
 4372
-2198
 2174 Answer
```

Let's try one more.

```
 72,354
-47,829
```

We can subtract 47,829 from either 70,000 or from 50,000. Let's try both.

```
    +1
 69,999
-47,829
```

Taking the 4 from 6, then each digit from 9 and the last from 10, we get 22,171.

```
     +1
 69,999
−47,829
 22,171
```

The calculation is now 22,171 + 2,354 (which is the amount over 70,000). It is not hard to add from left to right to get our answer of 24,525.

```
 72,354
−47,829
 24,525 Answer
```

Now, let's try subtracting from 50,000 and adding 22,354.

```
     +1
 49,999
−47,829
  2171
```

We now add 2,171 and 22,354 (which is the amount over 50,000). Twenty-two thousand plus 2,000 is 24,000. We now add 171 and 354. (To calculate mentally, I would add 354 plus 200, subtract 30 and then add 1. So, 354 + 200 is 554, then subtract 30 to get 524 and add the 1 to get our answer, 525.) Add 525 to 24,000 for the final answer of 24,525.

This method works best when either number is close to the number you choose to subtract from, so you end up with an easy addition. For instance, 421 − 381 or 6,123 − 5,789. Try them both.

Test yourself

Try these for yourself:

(a)   421
      −381

(b)   6,123
      −5,789

The answers are (a) 40 (b) 334

Checking subtraction by casting nines

For subtraction, the method used to check answers is similar to that used for addition, but with a small difference. Let us try an example.

```
 8465
-3896
 4569
```

Is the answer correct? Let's cast out the nines and see.

```
 8465   5
-3896   8
 4569   6
```

Five minus 8 equals 6? Can that be right? Obviously not. Although in the actual problem we are subtracting a smaller number from a larger number, with the substitutes, the number we are subtracting is larger.

We have two options. One is to add 9 to the number we are subtracting from. Five plus 9 equals 14. Then the problem reads:

$14 - 8 = 6$

This answer is correct, so our calculation was correct.

Here is the option I prefer, however. Call out the problem backwards as an addition. This is probably how you were taught to check subtractions in school. You add the answer to the number you subtracted to get your original number as your check answer.

Adding the substitutes upwards, we get $6 + 8 = 14$.

$6 + 8 = 14$

Adding the digits we get $1 + 4 = 5$, so our calculation checks.

$1 + 4 = 5$

Our answer is correct.

I set it out like this:

```
  8465   5
 −3896   8 +
  4569   6
```

Test yourself

Check these calculations for yourself to see if there are any mistakes. Cast out the nines to find any errors. If there is a mistake, correct it and then check your answer.

(a) 5,672   (b) 8,542   (c) 5,967   (d) 3,694
    −2,596   −1,495   −3,758   −1,236
     3,076   7,147   2,209   2,458

They were all right except for (b). Did you correct (b) and then check your answer by casting the nines? The correct answer is 7,047.

This method will find most mistakes in addition and subtraction. Make it part of your calculations. It only takes a moment and you will earn an enviable reputation for accuracy.

Note to parents and teachers

The difference is not so much with students' brains but with the methods they use. Teach the strugglers the methods the high achievers use and they can become high achievers too.

Chapter 14
Calculating left to right

Imagine you have $42.35 in your wallet. Which digit in the number is more important, the 4 or the 5? You don't say, 'I've got 5 cents in my wallet plus some dollars.' You are likely to say you have 40 dollars plus some change. The digit to the left of any number is the most important. So, why should we calculate beginning with the least significant digit; that is, calculating from right to left?

If you wanted to calculate how much you would pay for two items at $42.35 each, you would estimate the cost by doubling the $40 first to get $80. You would then double the 2 and then the 35 cents last. It makes sense to calculate from left to right because you are working with the most significant digits first.

And, it is easier to multiply $42.35 by 2 than to multiply 4,235 by 2. They are the same numbers, but changing the number into dollars and cents makes the calculation easier to handle. It is much easier to calculate $42.35 × 2 mentally than to calculate 4,235 × 2. Try it.

It is easy to calculate from left to right; it is just different to the way we have been taught.

Addition

How would you add 45 plus 37 in your head? Forty plus 30 is 70 and 5 plus 7 is 12, added to 70 is 82. Alternatively, 37 is 3 less than 40, so you could add 40 and subtract 3 to get the same answer.

Adding numbers from left to right is not difficult and is definitely the easier way to add mentally. Most people believe it is too hard so they never try.

To add 432 plus 237 in your head, you would begin with 432 and add 200 first, then add 30 and then the 7. Or, instead of adding 30 and 7, you could add 40 and subtract 3. You would say to yourself, '632, 672, 669.' To add 45 plus 37, it is easy to add 40 and subtract 3. You would say, '45, 85, 82.'

 Test yourself

Try these for yourself:

(a)   33 + 46 =   (b)   62 + 48 =

(c)   36 + 28 =   (d)   53 + 24 =

(e)   76 + 41 =   (f)   1,231 + 428 =

(continued)

Answers:

(a) 79   (b) 110
(c) 64   (d) 77
(e) 117   (f) 1,659

To calculate (f), I would add 400 to 1,231, then add 20 and then add 8.

Adding a series of numbers

How about adding a series of numbers? Let's add the following numbers.

```
  43
  25
  33
+27
```

Add the left-hand column first. Just remember though that this is the tens column.

$4 + 2 + 3 + 2 = 11$

Because this is the tens column, the answer is 110. Now you can add the units column.

I would add 10 first (7+3) to make 120, and then add the 5 and the 3 to get the answer of 128.

The calculation doesn't have to be mental. You can write the total of each column and then add the carried numbers. The calculation would then look like this:

```
  43
  25
  33
+27
  11
  18
 128 Answer
```

As you practise, you will find it is easy to do the whole calculation in your head.

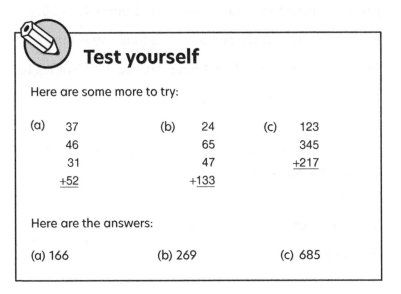

Test yourself

Here are some more to try:

(a)
```
  37
  46
  31
+52
```

(b)
```
  24
  65
  47
+133
```

(c)
```
 123
 345
+217
```

Here are the answers:

(a) 166   (b) 269   (c) 685

Multiplication

Now let's try multiplication. Here are some examples to get the idea.

To multiply 34 by 2, you would multiply 30 by 2 and then add 2 times 4. Two times 30 is 60, plus 2 times 4 (8) gives us 68. That was no more difficult than multiplying from right to left.

It makes sense to multiply the tens digit before the units digit. It is easier to keep track of your calculation and it also gives you an approximate answer when you are only partway into the calculation.

Just remember that 2 times 4 is 8, so 2 times 40 is 80.

In the same way, 2 times 3 is 6, so 2 times 30 is 60.

Just keep in mind that 40 is 4 times 10, and 30 is 3 times 10.

Now let's try some easier examples.

$2 \times 43 =$

You double the 40 first to get 80 and then add double the 3 to get our answer of 86.

$2 \times 47 =$

Two times 40 is 80.

Two times 7 is 14.

$80 + 14 = 94$

These calculations are not difficult.

How about 6 × 78?

Six times 70 is 420.

Six times 8 is 48. Add the 40 and then the 8.

$420 + 48 = 468$

You could also have calculated $6 \times (80 - 2)$:

$6 \times 80 = 480$

$6 \times 2 = 12$

$480 - 12 = 468$

Neither calculation is difficult. You decide which method you prefer.

Now, try calculating 312×2.

Begin with the 3, which is 300, and double it to get 600.

Now double the 12 to get 24. So, $2 \times 312 = 624$. Easy! You can practically call out the answer while you look at the numbers.

Test yourself

Try multiplying these numbers:

(a) $2 \times 43 =$

(b) $5 \times 78 =$

(c) $3 \times 34 =$

(d) $7 \times 32 =$

(e) $9 \times 27 =$

(f) $312 \times 4 =$

(continued)

Here are the answers:

(a) 86   (b) 390
(c) 102   (d) 224
(e) 243   (f) 1,248

Here is how you would calculate each of the problems.

$2 \times 43 = 80 + 6 = 86$

$5 \times 78 = 350 + 40 = 390$

$3 \times 34 = 90 + 12 = 102$

$7 \times 32 = 210 + 14 = 224$

$9 \times 27 = 180 + 63 = 243$

$312 \times 4 = 1,200 + 48 = 1,248$

Let's try one more.

$435 \times 4 = 1,600 + 140 = 1,740$

It looks impressive if you make the calculation in your head, but it is not difficult. You could easily multiply 4 times 435 in your head, but you wouldn't try it multiplying from right to left.

Using different reference numbers

This gives us freedom to use different reference numbers when we multiply.

Let's look at how we can multiply 73 × 78.

It is easy to use either 70 or 80 as our reference number. We will try both. First, let's try 70.

73 × 78 =

Add crossways.

78 + 3 = 81

We now multiply by the reference number, 70.

70 × 81 = 5,670

Multiply the numbers in the circles.

3 × 8 = 24

Add the two numbers to get the answer.

5,670 + 24 = 5,694

Here is the full calculation:

73 × 78 = ~~81~~

5,670

+   24

5,694   *Answer*

Now try using 80 as the reference number.

73 × 78 =

Subtract crossways.

73 − 2 = 71

Multiply by the reference number.

$80 \times 71 = 5,680$

Multiply the numbers in the circles.

$7 \times 2 = 14$

Add the two numbers to get the answer.

$5,680 + 14 = 5,694$

Here is the full calculation:

$$\begin{array}{ccccc} (80) & 73 & \times & 78 & = & \cancel{71} \\ & (-7) & & (-2) & & 5,680 \\ & & & & + & \underline{14} \\ & & & & & 5,694 \quad \textit{Answer} \end{array}$$

Let's look at how we can multiply 62 × 73.

We will use 60 as our reference number.

$$\begin{array}{ccccc} & (+2) & & (+13) & \\ (60) & 62 & \times & 73 & = \end{array}$$

Add crossways.

$73 + 2 = 75$

Multiply by the reference number.

$75 \times 60 =$

We can do this in two parts:

$70 \times 60 = 4,200$

$5 \times 60 = 300$

$4,200 + 300 = 4,500$

Now we multiply the numbers in the circles.

$2 \times 13 = 26$

Add the two numbers to get the answer.

$4,500 + 26 = 4,526$

$$\begin{array}{ccccc} & \textcircled{+2} & & \textcircled{+13} & \\ \textcircled{60} & 62 & \times & 73 & = & \cancel{75} \\ & & & & 4{,}500 \\ & & & & + \underline{\quad 26} \\ & & & & 4{,}526 \quad \textit{Answer} \end{array}$$

Left to right multiplication is not difficult. Now, instead of saying to yourself, 'How can I solve this problem?', you can say, 'What is the easiest way to solve it?'

Subtraction

Many people don't like subtraction because it can have some complicated carrying and borrowing. Would calculating from left to right make it even more difficult?

$54 - 31 = 23$

Fifty minus 30 is 20; 4 minus 1 is 3.

That was easy because there was no carrying or borrowing. Let's try some more.

$45 - 23 =$

Forty minus 20 leaves 20; 5 minus 3 is 2. The answer is 22. There was nothing to borrow or carry.

86 − 31 =

Eighty minus 30 is 50; 6 minus 1 is 5. The answer is 55.

So, what do we do if we have a calculation like 54 minus 37? In this case we would carry and borrow if we were doing the calculation with pen and paper.

Instead of subtracting 30 and then 7, I would round off upwards to 40. Forty is 3 more than 37, so I would subtract 40 and then give back the 3 that was more than the 37 we should have subtracted.

54 − 37 =

54 − 40 = 14

Now add the 3.

14 + 3 = 17 *Answer*

Let's try some more.

82 − 48 =

Round up to 50.

82 − 50 = 32

We rounded up by 2 so we add it back to get the answer.

32 + 2 = 34 *Answer*

87 − 29 =

Subtract 30 instead of 29 and then give back the extra 1.

87 − 30 = 57

57 + 1 = 58 *Answer*

Now let's try some three-digit numbers.

476 − 149 =

Four hundred minus 100 equals 300.

Round off 49 to 50 and the rest is easy.

$76 - 50 = 26$

Add 26 to 300, plus the extra 1 we subtracted, for our answer of 327.

There is no law that says you have to calculate from left to right. Neither is there a law that says you have to calculate from right to left. It is your choice.

You will find that adding and multiplying from left to right makes more sense and is easier to calculate mentally. It is easier to keep track of the numbers as you go.

It makes sense to calculate the important values first before calculating the minor values.

It's not as difficult as it might seem. It's easy with practice.

Chapter 15
Simple division

When do you need to be able to divide? You need to be able to do short division in your head when you are shopping at the supermarket, when you are watching a sporting event, when you are handling money or splitting up food.

If you have $30 to divide among 6 people, you need to be able to divide 6 into 30 to find out how much money each person should receive. You would find they receive $5 each.

$30 \div 6 = 5$

$5 \times 6 = 30$

If you are watching a cricket match and a team has scored 32 runs in 8 overs, how many runs per over have they scored? What is their average run rate? You need to divide 32 by 8 to find the answer. The answer is 4, or 4 runs per over.

$32 \div 8 = 4$

$4 \times 8 = 32$

You need to be able to divide so you know when you are getting the most for your money. Which is better value, a 2-litre bottle of drink for $2.75 or a 1.25-litre bottle for $2.00? You need to be able to divide to work out which is the better buy.

Many people think division is difficult and would rather not have to do it, but it is really not that hard. I will show you how to make division easy. Even if you are confident with simple division, it might be worthwhile reading this chapter.

Dividing smaller numbers

If you had to divide 10 lollies among 5 people, they would receive 2 lollies each. Ten can be divided evenly by 5.

If you divided 33 maths books among 4 people, they would each receive 8 books, and there would be 1 book left over. Thirty-three cannot be evenly divided by 4. We call the 1 book left over the remainder. We would write the calculation like this:

$$4 \,\vert\, \underline{33}$$
$$8 \;\; r1$$

Or like this:

$$8 \;\; r1$$
$$4 \,\vert\, \overline{33}$$

We could ask, what do we multiply by 4 to get an answer of 33, or as close to 33 as we can without going above? Four times 8 is 32, so the answer is 8. We subtract 32 from the number we are dividing to find the remainder (what is left over).

Dividing larger numbers

Here is how we would divide a larger number. To divide 3,721 by 4, we would set out the problem like this:

$$4 \overline{\smash{\big)}\, 3{,}721}$$

Or like this:

$$4 \,\overline{\big|\ 3{,}721}$$

We begin from the left-hand side of the number we are dividing. Three is the first digit on the left. We begin by asking, what do you multiply by 4 to get an answer of 3?

Three is less than 4 so we can't evenly divide 3 by 4. So we join the 3 to the next digit, 7, to make 37. What do we multiply by 4 to get an answer of 37? There is no whole number that gives you 37 when you multiply it by 4. We now ask, what will give an answer just below 37? The answer is 9, because $9 \times 4 = 36$. That is as close to 37 as we can get without going above. So, the answer is 9 ($9 \times 4 = 36$), with 1 left over to make 37. One is our remainder. We would write '9' above the 7 in 37 (or below, depending on how you set out the problem). The 1 left over is carried to the next digit and put in front of it. The 1 carried changes the next number from 2 to 12.

The calculation now looks like this:

$$4 \,\overline{\big|\ 3{,}7^{1}21} \atop \ \ 9$$

Or like this:

$$4 \,\overline{\big|\ 37^{1}21} \atop \ \ \ 9$$

We now divide 4 into 12. What number multiplied by 4 gives an answer of 12? The answer is 3 (3 × 4 = 12). Write 3 above (or below) the 2. There is no remainder as 3 times 4 is exactly 12.

The last digit is less than 4 so it can't be divided. Four divides into 1 zero times with 1 remainder.

The finished problem should look like this:

$$4 \overline{\smash{)}\, 37^\text{1}21}$$
$$9\ 30\ \ \text{r1}$$

Or:

$$9\ 30\ \ \text{r1}$$
$$4 \overline{\smash{)}\, 37^\text{1}21}$$

The 1 remainder can be expressed as a fraction, ¼. The ¼ comes from the 1 remainder over the divisor, 4. The answer would be 930¼, or 930.25.

This is a simple method and should be carried out on one line.

It is easy to calculate these problems in your head this way.

Dividing numbers with decimals

How would we divide 567.8 by 3?

We set out the problem in the usual way.

$$3 \overline{\smash{)}\, 567.8}$$

The calculation begins as usual.

Three divides once into 5 with 2 remainder. We carry the 2 remainder to the next digit, making 26. We write the answer, 1, below (or above) the 5 we divided. The calculation looks like this:

$$3 \mid \overline{5^267.8}$$
$$1$$

We now divide 26 by 3. Eight times 3 is 24, so the next digit of the answer is 8, with 2 remainder. We carry the 2 to the 7.

$$3 \mid \overline{5^26^27.8}$$
$$1\,8$$

Three divides into 27 exactly 9 times ($9 \times 3 = 27$), so the next digit of the answer is 9.

$$3 \mid \overline{5^26^27.8}$$
$$1\,8\,9$$

Because the decimal point follows the 7 in the number we are dividing, it will follow the digit in the answer below the 7.

We continue as before. Three divides into 8 two times with 2 remainder. Two is the next digit of the answer. We carry the remainder, 2, to the next digit.

$$3 \mid \overline{5^26^27.8^2}$$
$$1\,8\,9.2$$

Because there is no next digit, we must supply a digit ourselves. We can write a whole string of zeros after the last digit following a decimal point without changing the number.

We will calculate our answer to two decimal places, so we must make another division to see how we round off our answer. We write two more zeros to make three digits after the decimal.

$$3 \;|\; 5^2 6^2 7.8^2 00$$
$$\overline{1\,8\,9.2\,6}$$

Three divides into 20 six times ($6 \times 3 = 18$), with 2 remainder. The remainder is carried to the next digit, making 20 again. Because we will keep ending up with 2 remainder, you can see that this will go on forever. Three divides into 20 six times with 2 remainder, and this will continue infinitely.

$$3 \;|\; 5^2 6^2 7.8^2 0^2 0$$
$$\overline{1\,8\,9.2\,6\,6}$$

Because we are calculating to two decimal places, we have to decide whether to round off upwards or downwards. If the next digit after our required number of decimal places is 5 or above, we round off upwards; if the next digit is below 5 we round down. The third digit is 6 so we round the second digit off to 7. Our answer is 189.27.

Simple division using circles

Just as our method with circles can be used to multiply numbers easily, it can also be used in reverse for division. The method works best for division by 7, 8 and 9. I think it is easier to use the short division method we have just looked at, but you can try using the circles if you are still not sure of your tables.

Let's try a simple example, 56 ÷ 8:

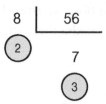

Here is how it works. We are dividing 56 by 8. We set the problem out as above, or, if you prefer, you can set the problem out like below. Stick to the way you have been taught.

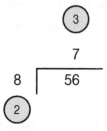

I will explain using the first layout. We draw a circle below the 8 (the number we are dividing by — the divisor) and then ask, how many do we need to make 10? The answer is 2, so we write 2 in a circle below the 8. We add the 2 to the tens digit of the number we are dividing (5 is the tens digit of 56) and get an answer of 7. Write 7 below the 6 in 56. Draw a circle below our answer (7). Again, how many more do we need to make 10? The answer is 3, so write 3 in the circle below the 7. Now multiply the numbers in the circles.

$2 \times 3 = 6$

Subtract 6 from the units digit of 56 to get the remainder.

6 − 6 = 0

There is 0 remainder. The answer is 7 with 0 remainder. Here is another example: 75 ÷ 9.

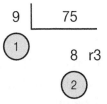

Nine is 1 below 10, so we write 1 in the circle below the 9. Add the 1 to the tens digit (7) to get an answer of 8. Write 8 as the answer below the 5. Draw a circle below the 8. How many more to make 10? The answer is 2. Write 2 in the circle below the 8. Multiply the numbers in the circles, 1 × 2, to get 2. Take 2 from the units digit (5) to get the remainder, 3. The answer is 8 r3.

Here is another example that will explain what we do when the result is too high.

$$\begin{array}{r} 7 \\ 8\,\overline{)\,52} \\ \textcircled{2} \end{array}$$

Eight is 2 below 10, so we write 2 in the circle. Two plus 5 equals 7. We write 7 above the units digit. We now draw another circle above the 7. How many to make 10? The answer is 3, so we write 3 in the circle. To get the remainder, we multiply the two numbers in the circles and take the answer from the units digit. Our work should look like this:

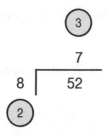

We find, though, that we can't take 6 from the units digit, 2. Our answer is too high. To rectify this, we drop the answer by 1 to 6, and write a small 1 in front of the units digit, 2, making it 12. Six is 4 below 10, so we write 4 in the circle.

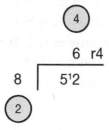

We multiply the two circled numbers, 2 × 4 = 8. We take 8 from the units digit, now 12; 12 − 8 = 4. Four is the remainder.

The answer is 6 r4.

Test yourself

Try these problems for yourself:

(a) 76 ÷ 9 =   (b) 76 ÷ 8 =

(c) 71 ÷ 8 =   (d) 62 ÷ 8 =

(e) 45 ÷ 7 =   (f) 57 ÷ 9 =

The answers are:

(a) 8 r4   (b) 9 r4   (c) 8 r7
(d) 7 r6   (e) 6 r3   (f) 6 r3

This method is useful if you are still learning your multiplication tables and have difficulty with division, or if you are not certain and just want to check your answer. As you get to know your tables better you will find standard short division to be easy. Next time you watch a sporting event, use these methods to see how your team is going.

Remainders

Let's go back to our problem at the beginning of the chapter. How would we divide 33 maths books among 4 students? You couldn't really say that each student is given 8.25 or 8¼ books each, unless you want to destroy one of the books! Each student receives 8 books and there is 1 book left over. You can then decide what to do with the extra book. We would write the answer as 8 r1, not 8.25 or 8¼.

If we were dividing up money, we could write the answer as 8.25, because this is 8 dollars and 25 cents.

Some problems in division require a whole remainder to make sense, others need the remainder expressed as a decimal.

Simple division by adding

Here is a simple method to divide any size number by 7, 8 or 9. Try it and you will find it easy and fun. Anyone can use this method, even if they don't know their multiplication tables well.

Division by nine: Two-digit numbers

There is an easy short cut for division by 9. Nine divides into 10 one time with 1 remainder. When you divide any number by nine you get an answer that equals the tens digit with a remainder of one for each ten. For instance, if you divide 9 into 30 it will go 3 times, with a remainder of 3. If you had $30 in your pocket, you could buy 3 books for $9 and get $1 change for each book you bought. If you divide 9 into 32 it will go three times, with a remainder of 3 plus the units digit, making a remainder of 5. So if you divide a two-digit number by 9, the first digit of the number is the answer and adding the digits of the number gives you the remainder. For instance, dividing 42 by 9, the first digit, 4, is the answer, and the sum of the digits, 4 + 2, is the remainder.

Here are some other examples:

$$42 \div 9 = 4 \text{ r}6$$
$$61 \div 9 = 6 \text{ r}7$$
$$23 \div 9 = 2 \text{ r}5$$

These are easy.

What do we do if the sum of the digits is 9 or higher? For instance, if we divide 65 by 9, the first digit, 6, is our answer and the remainder is the sum of the digits, 6 + 5 = 11. Our answer is 6 r11. But that doesn't make sense, because you can't have a remainder larger than your divisor. Nine divides into 11 one more time, so we add 1 to the answer (6 + 1), and the 2 left over from 11 becomes the new remainder. So the answer becomes 7 r2.

We could also have applied our shortcut to the 11 remainder. The first digit is 1, which we add to our answer, and the sum (1 + 1) gives a remainder of 2.

Let's look at another example.

$75 \div 9 =$

The first digit is 7.

Add the answer, 7, to the second digit, 5.

$7 + 5 = 12$

We can't have a remainder that is greater than the divisor, so we subtract 9 and add 1 more to our answer. The answer becomes 8 r3.

$75 \div 9 = 8$ r3 *Answer*

How about this example:

$58 \div 9 =$

The first digit is 5.

Add the answer, 5, to the second digit, 8.

$5 + 8 = 13$

Our answer is greater than the divisor, so we subtract 9 and add 1 more to our answer. The answer becomes 6 r4.

758 ÷ 9 = 6 r4 *Answer*

Test yourself

Try the shortcut with the following:

(a)   25 ÷ 9 =   (b)   61 ÷ 9 =

(c)   34 ÷ 9 =   (d)   75 ÷ 9 =

(e)   82 ÷ 9 =

Here are the answers:

(a) 2 r7   (b) 6 r7   (c) 3 r7

(d) 8 r3   (e) 9 r1

Division by nine: Three-digit numbers

Let's see how this works with longer numbers. The method remains the same.

Consider this example.

123 ÷ 9 =

The first digit of the number we are dividing is 1; this is the first digit of the answer.

Add the first digit of the answer to the next digit.

1 + 2 = 3

Remember when using this method that the final digit will make up the remainder.

Add the last digit of the answer to the next digit.

3 + 3 = 6

This number is the remainder.

13 r6 *Answer*

Let's try another example.

567 ÷ 9 =

The first digit of the number (5) is the first digit of the answer.

567 ÷ 9 =

5

Add the first digit of the answer (5) to the next digit of the number we are dividing.

5 + 6 = 11

Our answer is greater than the divisor, so we subtract 9 and add 1 to the previous digit of the answer and write 2 as the next digit of the answer.

567 ÷ 9 =

52

1

Add the last digit of the answer (2) to the next digit of the number we are dividing (7).

$2 + 7 = 9$

This number would be our remainder, but we can't have a remainder of 9 when we are dividing by 9 so we add 1 to the previous digit of the answer.

$567 \div 9 =$

52

11

Adding the carried numbers gives an answer of 63.

Test yourself

Try out these divisions by 9:

(a) $123 \div 9 =$ (b) $456 \div 9 =$

(c) $789 \div 9 =$ (d) $173 \div 9 =$

The answers are:

(a) 13 r6   (b) 50 r6
(c) 87 r6   (d) 19 r2

Division by nine: Any-length numbers

Let's try dividing some longer numbers by 9.

12,345 ÷ 9 =

The first digit of the number we are dividing is 1; this is the first digit of the answer.

Add the first digit of the answer to the next digit.

1 + 2 = 3

This is the next digit of the answer.

12,345 ÷ 9 = 13

Add the last digit of the answer to the next digit of the number to be divided.

3 + 3 = 6

12,345 ÷ 9 = 136

We repeat the procedure for the next digit of the answer.

6 + 4 = 10

Ten is greater than 9, so we subtract 9, add 1 to the previous digit of the answer and write the 1 remainder as the next digit of the answer.

12,345 ÷ 9 = 1371

Add the last digit of the answer to the next digit of the number to be divided.

1 + 5 = 6

1,371 r6

12,345 ÷ 9 = 1,371 r6 *Answer*

Test yourself

Try these for yourself:

(a)   51,235 ÷ 9 =

(b)   2,468 ÷ 9 =

(c)   72,951 ÷ 9 =

(d)   3,162 ÷ 9 =

Here are the answers:

(a) 5,692 r7

(b) 274 r2

(c) 8,105 r6

(d) 351 r3

Division by eight: Two-digit numbers

Dividing by 8 is exactly the same as dividing by 9 except you double each digit of the answer before adding it to the next digit of the number being divided. Let's try it, first with two-digit numbers.

$32 \div 8 =$

The first digit is the first digit of the answer. The first digit is 3, so 3 is the first digit of the answer.

$32 \div 8 =$

3

Now double 3 and add to the next digit.

$2 \times 3 = 6$

$6 + 2 = 8$

Seeing 8 is the divisor we can't have a remainder of 8 so we add 1 to the previous digit. The answer now becomes 4.

Test yourself

Try these for yourself:

(a)   64 ÷ 8 =

(b)   47 ÷ 8 =

(c)   35 ÷ 8 =

(d)   29 ÷ 8 =

Here are the answers:

(a) 8

(b) 5 r7

(c) 4 r3

(d) 3 r5

Division by eight: Three-digit numbers

Dividing longer numbers by 8 follows the same procedure. We double each digit of the answer and add it to the next digit of the number we are dividing.

Consider this example:

$123 \div 8 =$

The first digit of the number is the first digit of the answer.

$123 \div 8 =$

1

Double the 1 and add to the next digit.

$2 \times 1 = 2$
$2 + 2 = 4$
$123 \div 8 =$
14

Double the 4 ($2 \times 4 = 8$) and add to 3.

$8 + 3 = 11$

Eleven is greater than the divisor, 8, so we subtract 8 and add 1 to the previous digit, and what is left over is the remainder.

$123 \div 8 =$
14 r3
$+1$

This gives us a final answer of 15 r3.

$123 \div 8 =$ **15 r3** *Answer*

Test yourself

Try these for yourself:

(a) $312 \div 8 =$ (b) $243 \div 8 =$

(c) $534 \div 8 =$ (d) $745 \div 8 =$

Here are the answers:

(a) 39   (b) 30 r3
(c) 66 r6   (d) 93 r1

Division by eight: Any-length numbers

Let's try dividing some longer numbers by 8.

12,345 ÷ 8 =

The first digit of the answer is 1.

Double the 1 and add the next digit.

$2 \times 1 = 2$

$2 + 2 = 4$

This is going the same way as the previous example.

12,345 ÷ 8 =

14

Double the 4 and add the next digit.

$4 \times 2 = 8$

$8 + 3 = 11$

Eleven is greater than the divisor, 8, so subtract 8 and add 1 to the previous digit of the answer, and write 3 as the next digit.

12,345 ÷ 8 =

143

+1

Double 3 to get 6 and add to 4.

$6 + 4 = 10$

Ten is greater than the divisor, so subtract 8 and add to 3 and write 2 as the next digit of the answer.

$12{,}345 \div 8 =$

1,432

+11

Double 2 and add to 5.

$4 + 5 = 9$

Nine is greater than the divisor, so subtract 8 and write 1 as the remainder.

$12{,}345 \div 8 =$

1,432 r1

+111

Add the carried numbers for your answer.

$12{,}345 \div 8 =$ **1,543 r1** *Answer*

Test yourself

Try these for yourself:

(a) $3{,}521 \div 8 =$ (b) $5{,}217 \div 8 =$

(c) $35{,}821 \div 8 =$

Here are the answers:

(a) 440 r1   (b) 652 r1   (c) 4,477 r5

Division by seven: Two-digit numbers

The first digit of the dividend is the first digit of the answer. Triple each digit of the answer and add to the next digit. We triple each digit of the answer because we are, in effect, dividing by 10 and giving 3 remainder.

For instance, how many times will 7 divide into 20? If we divide 20 by 10 we get an answer of 2. However, we want to divide by 7 so we get 3 remainder for each time 10 divides into 20. Two remainders of 3 gives us a remainder of 6. So, 7 divides into 20 two times with a remainder of 6.

Try this example:

$49 \div 7 =$

The first digit of the answer is 4.

$49 \div 7 =$

4

Triple the first digit of the answer and add the next digit.

$3 \times 4 = 12$

$12 + 9 = 21$

Seven divides 3 times into 21, so we add 3 to our answer.

$4 + 3 = 7$

$49 \div 7 =$

4

+3

$49 \div 7 = 7$ *Answer*

Test yourself

Try these for yourself:

(a)   24 ÷ 7 =   (b)   31 ÷ 7 =

(c)   57 ÷ 7 =   (d)   62 ÷ 7 =

Here are the answers:

(a) 3 r3   (b) 4 r3
(c) 8 r1   (d) 8 r6

Dividing by seven: Three-digit numbers and longer numbers

Let's try dividing some longer numbers by 7.

$123 \div 7 =$

The first digit of the number becomes the first digit of the answer.

$123 \div 7 =$

1

Multiply the 1 by 3 and add to the next digit.

$3 \times 1 = 3$

$3 + 2 = 5$

$123 \div 7 =$

15

Multiply 5 by 3 and add to the next digit.

$5 \times 3 = 15$

$15 + 3 = 18$

Seven divides into 18 two times (14) with a remainder of 4.

$123 \div 7 =$

15 r4

2

We add the carried 2 and our answer becomes 17 r4.

Test yourself

Try these for yourself:

(a) $111 \div 7 =$ (b) $1,122 \div 7 =$

(c) $231 \div 7 =$ (d) $3,125 \div 7 =$

Here are the answers:

(a) 15 r6   (b) 160 r2

(c) 33   (d) 446 r3

Why does it work?

Your favourite store has your favourite maths book on sale for only $9; how many copies can you buy if you have

$32 in your pocket? For each $10 you have, you can buy one book and have $1 left over. So for $32, you can buy three books. You hand over $30 and get $1 change for each book. That gives you $3 change plus the $2 in your pocket, which leaves you with a total of $5. So, 32 divided by 9 gives you 3 with 5 remainder.

If the books were reduced to $8, you would get $2 change for each $10 you pay for the books. So, if you have $32, you would hand the assistant $30 and get $2 change for each book. That gives you $6 change, plus the $2 in your pocket, which would leave you with $8 — enough to buy another book.

Can you see why this method is so simple to use and why it works?

These methods are fun to play with and give insights into the way numbers work.

You don't need to know your multiplication tables to divide by 9.

You only need to be able to multiply by 2 to divide by 8, and you only need to be able to multiply by 3 to divide by 7.

Chapter 16
Long division by factors

Most people don't mind short division (or simple division) but they feel uneasy when it comes to long division. To divide a number by 6 you need to know your 6 times table. To divide a number by 7, you need to know your 7 times table. But what do you do if you want to divide a number by 36? Do you need to know your 36 times table? No, not if you divide using factors.

What are factors?

What are *factors*? We have already made use of factors when we used 20 as a reference number with our multiplication. To multiply by 20, we multiply by 2 and then by 10. Two times 10 equals 20. We are using factors, because 2 and 10 are factors of 20. Four and 5 are also factors of 20 because 4 times 5 equals 20.

Just as we made use of factors in multiplication, we can use factors to make long division easy. Let's try long division by 36.

What can we use as factors? Four times 9 is 36, and so is 6 times 6. We could also use 3 times 12. Let's try our calculation using 6 times 6.

We will use the following division as an example:

$2,340 \div 36 =$

We can set out the problem like this:

$$
\begin{array}{c|l}
6 & 2,340 \\
\hline
6 & \\
\hline
\end{array}
$$

Or like this:

$$
\begin{array}{c|l}
6 & \\
\hline
6 & 2,340 \\
\hline
\end{array}
$$

Use the layout that you are comfortable with.

Now, to get started we divide 2,340 by 6. We use the method we learnt in the previous chapter.

We begin by dividing the digit on the left. The digit on the left is 2, so we divide 6 into 2. Two is less than 6 so we can't divide 2 by 6, so we join 2 to the next digit, 3, to make 23.

What number do we multiply by 6 to get an answer of 23? There is no whole number that gives you 23 when you multiply it by 6. We now ask, what will give an answer just below 23? Three times 6 is 18. Four times six is 24, which is too high, so our answer is 3. Write 3 below the 3 of 23.

Subtract 18 (3 × 6) from 23 to get an answer of 5 for our remainder. We carry the 5 remainder to the next digit, 4, making it 54. Our work so far looks like this:

$$6 \mid 2,3^540$$
$$6 \mid 3$$

We now divide 6 into 54. What do we multiply by 6 to get an answer of 54? The answer is 9. Nine times 6 is exactly 54, so we write 9 and carry no remainder.

Now we have one digit left to divide. Zero divided by 6 is 0, so we write 0 as the final digit of the first answer. Our calculation looks like this:

$$6 \mid 2,3^540$$
$$6 \mid 3\ 90$$

Now we divide our answer, 390, by the second 6.

Six divides into 39 six times with 3 remainder (6 × 6 = 36). Write 6 below the 9 and carry the 3 remainder to make 30.

$$6 \mid 2,3^540$$
$$6 \mid 3\ 9^30$$
$$6$$

Six divides into 30 exactly 5 times, so the next digit of the answer is 5. Our answer is 65 with no remainder.

$$6 \mid 2,3^54\ 0$$
$$6 \mid 3\ 9^30$$
$$6\ 5$$

I have written the problem the way I was taught.

If you prefer the other layout your calculation would look like this:

$$
\begin{array}{r|l}
 & 6\,5 \\
\hline
6 & 3\ 9^30 \\
\hline
6 & 2{,}3^54\ 0 \\
\end{array}
$$

What are some other factors we could use? To divide by 48, we could divide by 6, then by 8 (6 × 8 = 48).

To divide by 25 we would divide by 5 twice (5 × 5 = 25).

To divide by 24 we could divide by 4, then by 6 (4 × 6 = 24). Or we could divide by 3, then by 8, or we could halve the number and then divide by 12.

A good general rule for dividing by factors is to divide by the smaller number first and then by the larger number. The idea is that you will have a smaller number to divide when it is time to divide by the larger number.

Dividing by numbers such as 14 and 16 should be easy to do mentally. It is easy to halve a number before dividing by a factor. If you had to divide 368 by 16 mentally, you would say, 'Half of 36 is 18, half of 8 is 4.' You have a subtotal of 184. It is easy to keep track of this as you divide by 8.

Eighteen divided by 8 is 2 with 2 remainder. The 2 carries to the final digit of the number, 4, giving 24. Twenty-four divided by 8 is exactly 3. The answer is 23 with no remainder. This can be easily done in your head.

If you had to divide 2,247 by 21, you would divide by 3 first, then by 7. By the time you divide by 7 you have a smaller number to work with.

$$2,247 \div 3 = 749$$
$$749 \div 7 = 107$$

It is easier to divide 749 by 7 than to divide 2,247 by 7.

Division by numbers ending in 5

To divide by a two-digit number ending in 5, double both numbers and use factors. As long as you double both numbers, the answer doesn't change. Think of 4 divided by 2. The answer is 2. Now double both numbers. It becomes 8 divided by 4. The answer remains the same. (This is why you can cancel fractions without changing the answer.)

Let's have a try:

$$1,120 \div 35 =$$

Double both numbers. Twice 11 is 22, and two times 20 is 40; so 1,120 doubled is 2,240. Thirty-five doubled is 70. The problem is now:

$$2,240 \div 70 =$$

To divide by 70, we divide by 10, then by 7. We are using factors.

$$2,240 \div 10 = 224$$
$$224 \div 7 = 32$$

This is an easy calculation. Seven divides into 22 three times ($3 \times 7 = 21$) with 1 remainder, and divides into 14 (1 carried) twice.

This is a useful short cut for division by 15, 25, 35 and 45. You can also use it for 55. This method also applies to division by 1.5, 2.5, 3.5, 4.5 and 5.5.

Let's try another:

512 ÷ 35 =

Five hundred doubled is 1,000. Twelve doubled is 24. So, 512 doubled is 1,024. Thirty-five doubled is 70.

The problem is now:

1,024 ÷ 70

Divide 1,024 by 10, then by 7.

1,024 ÷ 10 = 102.4
102.4 ÷ 7 =

Seven divides into 10 once; 1 is the first digit of the answer. Carry the 3 remainder to the 2, giving 32.

32 ÷ 7 = 4 r4

We now have an answer of 14 with a remainder. We carry the 4 to the next digit, 4, to get 44.

44 ÷ 7 = 6 r2

Our answer is $14\frac{2}{7}$.

We have to be careful with the remainder. The 2 remainder we obtained is not the remainder for the original problem. We will now look at obtaining a valid remainder when we divide using factors.

Finding a remainder

Sometimes when we divide, we would like a remainder instead of a decimal. How do we get a remainder when we divide using factors? We actually have two remainders during the calculation.

The rule is:

Multiply the first divisor by the second remainder and then add the first remainder.

For example:

34,567 ÷ 36

```
            960  r1 ←
  6 |  5,761  r1 ←  + 
  6 | 34,567
```

We begin by multiplying the corners

6 x 1 = 6

Then we add the first remainder, 1. The final remainder is 7, or $\frac{7}{36}$.

Test yourself

Try these for yourself, calculating the remainder:

(a)   2,345 ÷ 36 =   (b)   2,713 ÷ 25 =

The answers are:

(a) 65 r5   (b) 108 r13

Working with decimals

We can divide numbers using factors to as many decimal places as we like.

Put as many zeros after the decimal as the number of decimal places you require, and then add one more. This ensures that your final decimal place is accurate.

If you were dividing 1,486 by 28 and you need accuracy to two decimal places, put three zeros after the number. You would divide 1,486.000 by 28.

$$
\begin{array}{r}
53.071 \\
7 \enclose{longdiv}{371.500} \\
4 \enclose{longdiv}{1,486.000}
\end{array}
$$

Rounding off decimals

To round off to two decimal places, look at the third digit after the decimal. If it is below 5, you leave the second digit as it is. If the third digit is 5 or more, add 1 to the second digit.

In this case, the third digit after the decimal is 1. One is lower than 5 so we round off the answer by leaving the second digit as 7.

The answer, to two decimal places, is 53.07.

If we were rounding off to one decimal place the answer would be 53.1, as the second digit, 7, is higher than 5, so we round off upwards.

The answer to eight decimal places is 53.07142857.

To round off to seven decimal places, we look at the eighth digit, which is 7. This is above 5, so the 5 is rounded off upwards to 6. The answer to seven decimal places is 53.0714286.

To round off to six decimal places, we look at the seventh digit, which is a 5. If the next digit is 5 or greater we round off upwards, so the 8 is rounded off upwards to 9. The answer to six decimal places is 53.071429.

To round off to five decimal places, we look at the sixth digit, which is a 9, so the 2 is rounded off upwards to 3. The answer to five decimal places is 53.07143.

Test yourself

Try these for yourself. Calculate these to two decimal places:

(a)   4,166 ÷ 42 =   (Use 6 × 7 as factors)

(b)   2,545 ÷ 35 =   (Use 5 × 7 as factors)

(c)   4,213 ÷ 27 =   (Use 3 × 9 as factors)

(d)   7,817 ÷ 36 =   (Use 6 × 6 as factors)

The answers are:

(a) 99.19   (b) 72.71   (c) 156.04   (d) 217.14

Long division by factors allows you to do many mental calculations that most people would not attempt. I constantly calculate sporting statistics mentally while a game is in progress to check how my team is going. It is a fun way to practise the strategies.

If I want to calculate the runs per over during a cricket match I simply divide the score by the number of overs bowled. When more than 12 overs have been bowled I will use factors for my calculations.

Why not try these calculations with your favourite sports and hobbies?

Chapter 17
Standard long division made easy

In the previous chapter we saw how to divide by large numbers using factors. This principle is central to all long division, including standard long division commonly taught in schools.

Division by factors worked well for division by numbers such as 36 (6 × 6), 27 (3 × 9), and any other number that can be easily reduced to factors. But what about division by numbers such as 29, 31 or 37, that can't be reduced to factors? These numbers are called prime numbers; the only factors of a prime number are 1 and the number itself.

Let me explain how our method works in these cases.

If we want to divide a number like 12,345 by 29, this is how we do it. We can't use our long division by factors method because 29 is a prime number. It can't be broken up into factors, so we use standard long division.

We set the problem out like this:

$$29 \overline{)\ 12345}$$

We then proceed as we did for short division. We try to divide 29 into 1, which is the first digit of 12,345, and of course we can't do it. So we join the next digit and divide 29 into 12. We find that 12 is also too small — it is less than the number we are dividing by — so we join the next digit to get 123.

Now we divide 29 into 123. This is where we have a problem; most people don't know the 29 times table, so how can they know how many times 29 will divide into 123?

The method is easy. This is how everyone does long division but they don't always explain it this way. Firstly, we round off the number we are dividing by. We would round off 29 to 30. We divide by 30 as we go, to estimate the answer, and then we calculate for 29.

How do we divide by 30? Thirty is 10 times 3, so we divide by 10 and by 3 to estimate each digit of the answer. So, we divide 123 by 30 to get our estimate for the first digit of the answer. We divide 123 by 10 and then by 3. To roughly divide 123 by 10 we can simply drop the final digit of the number, so we drop the 3 from 123 to get 12. Now divide 12 by 3 to get an answer of 4. Write 4 above the 3 of 123. Our work looks like this:

$$29 \overline{)\ \overset{\ \ \ 4}{12345}}$$

Now we multiply 4 times 29 to find what the remainder will be. Four times 29 is 116. (An easy way to multiply 29 by 4 is to multiply 30 by 4 and then subtract 4.)

$$4 \times (30-1) = 120 - 4 = 116$$

Write 116 below 123 and subtract to find the remainder.

$$123 - 116 = 7$$

We bring down the next digit of the number we are dividing and write an X beneath to remind us the digit has been used.

```
        4
29 ⌐ 12345
    116X
     74
```

We now divide 74 by 29.

We divide 74 by 10 to get 7 (we just drop the last digit to get an approximate answer), and then divide by 3. Three divides into 7 twice, so 2 is the next digit of our answer.

Multiply 2 by 29 to get 58 (twice 30 minus 2), and then subtract from 74. The answer is 16. Then we bring down the 5, writing the X below.

```
       42
29 ⌐ 12345
    116XX
     74
     58
    165
```

We now divide 165 by 29.

Roughly dividing 169 by 10 we get 16. Sixteen divided by 3 is 5.

Multiply 29 by 5 to get 145. Subtract 145 from 165 for a remainder of 20.

```
           425
   29 | 12345
        116XX
          74
          58
         165
         145
          20r
```

Our answer is 425 with 20 remainder. The calculation is done. Our only real division was by 3.

Long division is easy if you regard the problem as an exercise with factors. Even though we are dividing by a prime number, we make our estimates by rounding off and using factors.

The general rule for standard long division is this:

Round off the divisor to the nearest ten, hundred or thousand to make an easy estimate.

If you are dividing by 31, round off to 30 and divide by 3 and 10.

If you are dividing by 87, round off to 90 and divide by 9 and 10.

If you are dividing by 321, round off to 300 and divide by 3 and 100.

If you are dividing by 487, round off to 500 and divide by 5 and 100.

If you are dividing by 6,142, round off to 6,000 and divide by 6 and 1,000.

This way, you are able to make an easy estimate and then proceed in the usual way.

Remember that these calculations are really only estimates. If you are dividing by 31 and you make your estimate by dividing by 30, the answer is not exact — it is only an approximation. For instance, what if you were dividing 241 by 31?

You round off to 30 and divide by 10 and by 3.

Two hundred and forty-one divided by 10 is 24. We simply drop the 1.

Twenty-four divided by 3 is 8.

We multiply 8 times 31 and subtract the answer from 241 to get our remainder.

Eight times 31 is 248. This is greater than the number we are dividing so we cannot subtract it. Our answer was too high. We drop to 7.

Seven times 31 is 217.

We subtract 217 from 241 for our remainder.

$$241 - 217 = 24$$

So 241 divided by 31 is 7 with 24 remainder.

We could have seen this by observing that 8 times 30 is 240 and 241 is only 1 more. So we can anticipate that 8 times 31 will be too high.

How about dividing 239 by 29?

We round off to 30 and divide by 10 and by 3 for our estimate.

Two hundred and thirty-nine divided by 10 is 23.

Twenty-three divided by 3 is 7.

But like the previous case, we can see the real answer is probably different.

Twenty-nine is less than 30 and 30 divides into 239 almost 8 times, as it is only 1 less than 240, which is exactly 8 times 30. So we multiply 8 times 29 to get 232, which is a valid answer.

Had we not seen this and chosen 7 as our answer, our remainder would have been too large.

Seven times 29 is 203. Subtract 203 to find the remainder.

$$239 - 203 = 36$$

This is higher than our divisor so we know we have to raise the last digit of our answer.

What if we are dividing by 252? What do we round it off to? Two hundred is too low and 300 is too high. The easy way would be to double both the number we are dividing and the divisor, which won't change the answer but will give us an easier calculation.

Let's try it.

$$2,233 \div 252 =$$

Doubling both numbers we get $4,466 \div 504$. We now divide by 500 (5 × 100) for our estimates and correct as we go.

```
            8 r434
504 | 4,466
      4,032
       434
```

We divide 4,466 by 500 for our estimate. We divide using factors of 100 × 5:

4,466 ÷ 100 gives 44

44 ÷ 5 gives 8 $(8 \times 5 = 40)$

Now, 8 × 504 = 4,032, subtracted from 4,466 gives a remainder of 434. If the remainder is important (we are not calculating the answer as a decimal), because we doubled the numbers we are working with, the remainder will be double the true remainder. That is because our remainder is a fraction of 504 and not 252. So, for a final step to find the remainder, we divide 434 by 2 to get an answer of 8 r217. If you are calculating the answer to any number of decimal places then no such change is necessary. That is because $^{217}/_{252}$ is the same as $^{434}/_{504}$.

 Test yourself

Try these problems for yourself. Calculate the remainder for a) and the answer to one decimal place for b).

(a)   2,456 ÷ 389 =   (b)   3,456 ÷ 151 =

The answers are:

(a) 6 r122   (b) 22.9

The answer to (b) is 22.8874 to four decimal places, 22.887 to three decimal places, 22.89 to two decimal places and

22.9 to one decimal place, which was the required answer. For (b) you would have doubled both numbers to make the calculation 6,912 ÷ 302. Then you estimate each digit of the answer by dividing by 100 × 3.

Long division is not difficult. It has a bad reputation. It doesn't deserve it. Many people have learnt long division, but they haven't been taught properly. Anything is difficult if you don't understand it and you don't know how to do it very well.

Break big problems down to little problems and you can do them.

In the next chapter you will learn an easy method to do long division in your head.

Note to parents and teachers

I was speaking to teachers at a special government programme in the United States and I said I always use factors for long division, even when I am dividing by a prime number. That was too much for one teacher, who was one of the organisers of the programme, and he challenged me to explain myself. I gave a quick summary of this chapter, and he said, 'You know, that is how I have always done long division, but I have never thought to explain it that way before.'

Chapter 18
Direct long division

Here is a very easy method for doing long division in your head. It is very similar to our method of short division. As many people have trouble doing long division, even with a pen and paper, if you master this method people will think you are a genius.

Here's how it works.

Let's say you want to divide 195 by 32.

Here is how I set out the problem:

32

(−2)

30 | 195

We round off 32 to 30 by subtracting 2. Then we divide both numbers by 10 by moving the decimal point one place to the left, making the number we are dividing 19.5

and the divisor 3. Now we divide by 3, which is easy, and we adjust for the −2 as we go.

Three won't divide into 1 so we divide the first two digits by 3. Three divides into 19 six times with 1 remainder (6 × 3 = 18). The 1 is carried to the next digit of the dividend (the number we are dividing), which makes 15.

```
    32
   (-2)
   3̶0̶ │  19.¹5
        ─────────
          6
```

Before we divide into 15 we adjust for the 2 in 32. We multiply the last digit of the answer, 6, by the units digit in the divisor, 32, which is 2.

$6 \times 2 = 12$

We subtract 12 from the working number, 15, to get 3.

What we have done is multiplied 32 by 6 and subtracted the answer to get 3 remainder. How did we do this?

We subtracted 6 times 32 by first subtracting 6 times 30, then 6 times 2. This is really a simple method for doing standard long division in your head.

```
    32
   (-2)
   3̶0̶ │  19.¹5
        ─────────
          6 r3
```

Let's try another one. Let's divide 430 by 32 to 2 decimal places. We set out the problem like this:

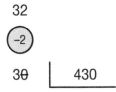

```
32
 -2
30  |  430
```

We divide both numbers by 10 to get 3 and 43. We add three zeros after the decimal so that we can calculate the answer to two decimal places. (Always add one more 0 than the number of decimal places required.) Our problem now looks like this:

```
32
 -2
30  |  43.000
```

We divide 3 into 4 for an answer of 1 with 1 remainder, which we carry to the 3, making 13. Then we multiply the answer, 1, by −2 to get an answer of −2. Take this 2 from our working number, 13, to get 11.

How many times will 3 divide into 11? Three times 3 is 9, so the answer is 3 times, with 2 remainder. The 2 remainder is carried to the next digit, making 20.

```
32
 -2
30  |  4¹3.²0
        1 3
```

We adjust the 20 by multiplying: $3 \times -2 = -6$. Then we subtract.

$20 - 6 = 14$

We now divide 14 by 3. Three divides into 14 four times, with 2 remainder. Four is the next digit of our answer.

```
        32

        (-2)

        30  │ 4¹3.²0²0
            ─────────
             1 3. 4
```

Multiply the last digit of the answer, 4, by the 2 of 32, to get 8. Then:

$20 - 8 = 12$

Three divides into 12 exactly 4 times with no remainder.

The next step would be to multiply 4 times 2 to get 8.

We subtract 8 from our remainder, which is 0, so we end up with a negative answer. That won't do, so we drop our last digit of the answer by 1. Our last digit of the answer is now 3, with 3 remainder.

```
        32

        (-2)

        30  │ 4¹3.²0²0³0
            ──────────
             1 3. 4 3
```

Three times 2 is 6, subtracted from 30 leaves 24. Twenty-four divided by 3 is 8 (with 0 remainder). We need a remainder to subtract from, so we drop the 8 to 7 with 3 remainder. The answer now becomes 13.437. Because we want an answer correct to two decimal places, we look at the third digit after the decimal. If the digit is 5 or higher we round off upwards; if the digit is less than 5 we round off downwards. In this case the digit is 7 so we round off the previous digit upwards. The answer is 13.44, correct to two decimal places.

Estimating answers

Let's divide 32 into 240. We set the problem out as before:

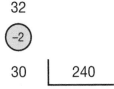

We begin by dividing both numbers by 10. Then, 3 divides into 24 eight times, with 0 remainder. The 0 remainder is carried to the next digit, 0, making a new working number of 0. We adjust for the 2 in 32 by multiplying the last digit of our answer by the 2 in 32 and subtracting from our working number, 0. Eight times 2 is 16. Zero minus 16 is −16. We can't work with −16 so the last digit of our answer was too high.

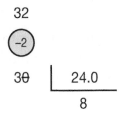

(We can easily see this, as 8 times 30 is exactly 240. We are not dividing by 30, but by 32. The answer is correct for 30, but clearly it is not correct for 32. Because each answer is only an estimate, we can see in this case we will have to drop our answer by 1.)

We change the answer to 7, so: $7 \times 3 = 21$, with 3 remainder.

Our calculation looks like this:

$$
\begin{array}{c}
32 \\
\boxed{-2} \\
3\!\!\!/0 \quad \big|\ 24.^30 \\
\hline
\quad\quad 7
\end{array}
$$

We now multiply:

$7 \times 2 = 14$

Then we subtract from 30:

$30 - 14 = 16$

The answer is 7 with 16 remainder.

If we want to take the answer to one decimal place, we now divide 16 by 3 for the next digit of the answer. Three divides into 16 five times, with 1 remainder. The 1 is carried to the next digit to make 10.

We multiply the last answer, 5, by the 2 in 32 to get an answer of 10. Subtract 10 from our working number 10 to get an answer of 0. There is no remainder. The answer is exactly 7.5.

The problem now looks like this:

```
 32
(-2)
 30 │ 24.³0¹0
    └─────────
       7. 5
```

This is not difficult, but you need to keep in mind that you have to allow for the subtraction from the remainder.

Reverse technique - rounding off upwards

Let's see how direct long division works for dividing by numbers with a high units digit, such as 39. We would round the 39 upwards to 40. For example, let's divide 231 by 39. We set the problem out like this:

```
 39
(+1)
 40 │ 231
```

We divide both numbers by 10, so 231 becomes 23.1 and 40 becomes 4. Four won't divide into 2, so we divide 4 into 23. Four divides 5 times into 23 with 3 remainder (4 × 5 = 20). We carry the 3 remainder to the next digit, 1, making 31. Our work looks like this:

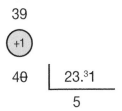

```
 39
(+1)
 40 │ 23.³1
    └────────
        5
```

We correct our remainder for 39 instead of 40 by adding 5 times the 1, which we added to make 40.

$5 \times 1 = 5$
$31 + 5 = 36$

Our answer is 5 with 36 remainder.

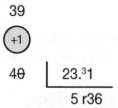

$$40 \enclose{longdiv}{23.^31}$$
$$5 \; r36$$

Remember, *these are only estimates*. In the following case we have a situation where the estimate is not correct, because 39 is less than 40. Let's say we want to divide 319 by 39. Again, we round off to 40. Our calculation would look like this:

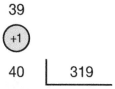

$$40 \enclose{longdiv}{319}$$

We divide both numbers by 10.

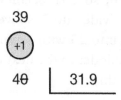

$$40 \enclose{longdiv}{31.9}$$

Four divides into 31 seven times, with 3 remainder. We write the 7 below the 1 in 31 and carry the 3 remainder to the next digit, 9, making 39.

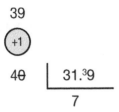

39

(+1)

4̶0̶ | 31.³9
            7

We multiply our answer, 7, by the +1 to get 7. We add 7 to 39, giving an answer of 46, which is our remainder. We can't have a remainder greater than our divisor, so, in this case, we have to raise the last digit of our answer by 1, to 8.

39

(+1)

4̶0̶ | 31.⁻¹9
            8

Four times 8 is 32, with −1 remainder, which we carry as −10, and we ignore the 9 for the moment. We multiply 8 times +1 to get +8, which we add to the next digit, 9, to get 17, then we minus the 10 carried to get a remainder of 7.

39

(+1)

4̶0̶ | 31.⁻¹9
           8 r7

With practice, this concept will become easy.

Test yourself

Try these problems for yourself. Calculate to one decimal place.

(a)   224 ÷ 29 =   (b)   224 ÷ 41 =

Let's calculate the answers together.

29

30 | 22.¹40
    7. 72

We have divided both numbers by 10. Three divides into 22 seven times (3 × 7 = 21) with 1 remainder, which we carry to the next digit, 4, to make 14. We multiply 7 × 1 = 7, and add to the 14 to get 21.

Now divide 21 by 3 to get 7, with no remainder to carry. Multiply the last digit of the answer, 7, by 1 to get 7, which we add to 0 for our next working number. Seven divided by 3 is 2, with 1 remainder, which we carry to the next digit, 0, to get 10. We don't need to go any further because we only need an answer to one decimal place; 7.72 rounds off to 7.7. This is our answer.

Now for the second problem

41

40 | 22.²4³0
    5. 4 6

We divide both numbers by 10. Four divides into 22 five times (5 × 4 = 20), with 2 remainder, which we carry to the next digit, 4, to make 24. Now multiply 5 (the last digit of the answer) by −1 to get −5. Subtract the 5 from 24 to get 19.

We now divide 19 by 4 to get an answer of 4 with 3 remainder. The 3 carried gives us a working number of 30. Now multiply the previous digit of the answer, 4, by −1 to get −4. Subtract 4 from 30 to get 26. Four divides six times into 26, and we can forget the remainder. Our answer so far is 5.46. We can stop now because we only need one decimal place. Rounded off to one decimal place we have 5.5 as our answer.

With practice, these calculations can easily be done in your head. Divide the last four digits of your telephone number by different numbers in your head for practice. Try some other numbers as well. You will find it is much easier than you imagine.

Chapter 19
Long division by adding

This method of division is great for dividing by numbers a little lower than powers of 10 or multiples of powers of 10. It also helps to illustrate what division is all about.

How many times will 9 divide into 10? One time with 1 remainder. That is, for every 10, 9 will divide once with 1 remainder. So, for 20, 9 will divide twice with 2 remainder. For 40, 9 will divide four times with 4 remainder. For every 10, 9 divides once with 1 remainder.

If you have a handful of dollars, you can buy one item costing 90 cents for each dollar and have 10 cents change. If you have enough dollars, you can buy some extra items out of the change. This brings us to a new and easy way of calculating some long division problems. If you are dividing by 90, divide by 100 (a dollar) and give back the change.

For instance, if you are buying a drink for 95 cents and you have $1.20, you would hand over a dollar and keep the 20 cents in your pocket, plus you would get 5 cents change, making 25 cents left over, or 25 cents remainder.

So, you could say, 95 divides into 120 one time with 25 remainder.

Dividing by two-digit numbers by adding

Let's try an example:

```
 100
 96 │  234
 (4)
```

We are dividing 234 by 96. We write the problem in the conventional way but we draw a circle below the divisor, 96, and write 4 in the circle. (How many to make 100?)

Now, instead of dividing by 96, we divide by 100. How many times will 100 divide into 234? Two times with 34 remainder. Write 2 as the answer.

We have our 34 remainder plus another 4 remainder for every 100. We have two hundreds, so our remainder is 2 times 4, which is 8, plus the 34 from 234, giving us a total of 42 remainder.

We would set out the problem like this:

```
 100      2
 96 │   234
 (4)     8+
        42  remainder
```

We don't add the hundreds digit because we have finished with it. In effect, what we are saying is that 100 divides into 234 twice, with 34 left over. Because we are really dividing by 96, we also have 4 remainder for each hundred it divides into.

If we were purchasing items for 96 cents and we had $2.34 in our pocket, we would hand $2.00 to the seller and keep the 34 cents in our pocket. We would get 8 cents change which, added to the 34 cents, makes a total of 42 cents.

Let's try another.

$705 \div 89 =$

We set it out:

```
       100
    89 |  705
   (11)
```

Eighty-nine is 11 less than 100 so we write 11 below the 89.

How many hundreds are there in 705? Obviously 7, so we write 7 as the answer.

What is our remainder? For every hundred, we have 11 remainder. We have 7 hundreds so we have 7 times 11 remainder. Seven times 11 is 77, plus the 5 left over from 705 gives us 82 (77 + 5 = 82).

```
       100      7
    89 |  705
   (11)       77
              82
```

Again, we add the 77 to the 5, not 705, as we have finished with the hundreds and are only concerned with the 5 remainder.

So, the answer is 705 ÷ 89 = 7 with 82 remainder.

Let's take the answer to two decimal places.

```
100     7.    _____
89 | 705.000
 11    77 ˣ_____
       820
```

We now divide 89 into 820.

We have 8 hundreds, so 8 is the next digit of the answer. Eight times 11 equals 88.

```
100     7.8   _____
89 | 705.000
 11    77 ˣ_____
       820
        88
```

Eighty-eight plus 20 equals 108 remainder. This is obviously too high as it is greater than our divisor, so we add 1 to our answer. Our calculation now looks like this:

```
100     7.9   _____
89 | 705.000
 11    77 ˣ_____
       820
        99
       +19
```

223

We now multiply our new answer (9) by 11 to get our new remainder. Because we had to increase the answer by one, we subtract one times the working divisor, 100, from our remainder. We cross out the hundreds digit of the remainder to get an actual remainder of 19.

We bring down the next zero to make it 190.

We can see that 89 will divide twice into 190 so we can simply write 2 as the next digit of the answer. (The '90' part of the number is already greater than our divisor.)

(If we couldn't see at a glance that the answer is 2, we could write 1 as the next digit because there is 1 hundred in 190. One times 11 gives us 11 to add to the 90 of 190 to make 101 remainder. Seeing our divisor is only 89, we can't have a remainder of 101.)

$$
\begin{array}{r}
100 \quad 7.92 \\
89\,\big|\,\overline{705.000} \\
\textcircled{11} \quad 77^{xx} \\
\hline
820 \\
99 \\
\hline
4{,}190 \\
22 \\
\hline
\cancel{1}12 \quad \text{remainder}
\end{array}
$$

Because we increased the digit of the answer by 1, we subtract 1 times 100 from the remainder. One hundred and twelve minus 100 is 12. Bring down the final zero to make 120.

Eighty-nine divides once into 120, which gives us the final digit of the answer — making the answer 7.921. Seeing we only need our answer to two decimal places, we can round off to 7.92. You could easily solve this entirely in your head.

Test yourself

Try calculating all of these in your head, just giving the answer with the remainder.

(a) 645 ÷ 98 =   (b) 2,345 ÷ 95 =

(c) 234 ÷ 88 =   (d) 1,234 ÷ 89 =

The answers are:

(a) 6 r57   (b) 24 r65   (c) 2 r58   (d) 13 r77

Easy, weren't they?

This method works well when dividing by numbers just below a power of 10, or a multiple of a power of 10, but it can be made to work for other numbers as well.

Dividing by three-digit numbers by adding

Dividing by three-digit numbers follows the same procedure. How many items costing $1.87 could you buy if you had $234.56 in your pocket? Let's try it.

Consider this example:

$23,456 \div 187 =$

We set out the problem like this:

```
    200
        ┌─────────
187 │ 23,456
 ⓐ13
```

We use a working divisor of 200 because 187 is 200 minus 13. We add the 13 to the calculation in a circle.

We make our first calculation. Two hundred divides into 234 one time, so the first digit of the answer is 1. We write the 1 over the 4.

We multiply the answer, 1, by the number in the circle, 13, to get an answer of 13. Write 13 under 234 and add it to the 34, which are the next two digits in the number we are dividing.

Consider you are buying an item for $187 and have $234 in your pocket. You hand over $200 and get $13 change. You add the change to the $34 in your pocket for a total of $47.

```
    34   +   13   =     47
    200        1
        ┌─────────
187 │ 23,456
 ⓐ13      13
             47
```

Now bring down the next digit, 5, to make 475.

Divide 475 by 200. Two hundred divides into 400 twice, so 2 is the next digit of the answer.

$$
\begin{array}{r}
200 \quad\; 12 \\
\hline
187 \;\big|\; 23{,}456 \\
\textcircled{13} \quad\; \underline{13^{\text{X}}} \\
475
\end{array}
$$

Multiply 2 by 13 to get 26. We add the 26 to our partial remainder of 75. (We divided 475 by 200 and got an answer of 2 with a remainder of 75.)

$75 + 26 = 101$

Bring down the next digit, 6.

$$
\begin{array}{r}
200 \quad\; 12 \\
\hline
187 \;\big|\; 23{,}456 \\
\textcircled{13} \quad\; \underline{13^{\text{XX}}} \\
475 \\
\underline{26} \\
1{,}016
\end{array}
$$

Divide 1,016 by 200. Two hundred divides into 1,000 five times, so the next digit of the answer is 5. Five times 13 is 65. Sixty-five plus 16 equals 81; this is our remainder.

$$
\begin{array}{r}
200 \qquad 125 \\
187\ \overline{)\ 23{,}456} \\
\textcircled{13} \quad \underline{13^{xx}} \\
475 \\
\underline{26\ \ } \\
1{,}016 \\
\underline{65\ } \\
81 \quad \text{remainder}
\end{array}
$$

The answer is 125 with 81 remainder (125 r81).

It is much easier to multiply the 13 by each digit of the answer than to multiply 187 by each digit.

The next example shows you what to do when you have a remainder when you divide by your working divisor.

$4{,}567 \div 293 =$

We set out the problem like this:

$$
\begin{array}{r}
300 \\
293\ \overline{)\ 4{,}567} \\
\textcircled{7}
\end{array}
$$

We divide 400 by 300 to get an answer of 1 with 100 remainder. I write a small 1 to indicate the remainder.

One times 7 is 7, so we add 7 to 56, plus the 100 carried to get our remainder, which gives us 163.

```
300      1
293 | 4,567
 7     ¹07
       163
```

Bring down the next digit of the number we are dividing. We now divide 1,637 by 300.

Three hundred divides into 1,600 five times with 100 remainder. We follow the same procedure.

Thirty-seven plus 135 gives us our remainder of 172. We had to carry a remainder twice in this calculation.

```
300      15
293 | 4,567
 7     ¹07ˣ
       1,637
        ¹35
       172   remainder
```

The answer is 15 with 172 remainder (15 r172).

Here is another example:

$45,678 \div 378 =$

Three hundred and seventy-eight is 22 less than 400 so we can easily use this method.

Here is how we set out the problem:

```
400
378 | 45,678
 22
```

We use 400 as our working divisor.

The first calculation is easy. Four hundred divides once into 456.

$$
\begin{array}{c|c}
400 & 1 \\
378 & 45{,}678 \\
\textcircled{22} & \underline{22} \\
& 78
\end{array}
$$

One times 22 is 22. Add 22 to 56 to get 78.

Bring down the next digit of the number we are dividing, 7.

$$
\begin{array}{c|c}
400 & 1 \\
378 & 45{,}678 \\
\textcircled{22} & \underline{22}^{\text{x}} \\
& 787
\end{array}
$$

Four hundred divides once into 787 but 787 is almost 800 and we are not actually dividing by 400, but by 378. We can see the answer is probably 2. Let's try it.

$$
\begin{array}{c|c}
400 & 12 \\
378 & 45{,}678 \\
\textcircled{22} & \underline{22}^{\text{x}} \\
& 787 \\
& \underline{44} \\
& 131
\end{array}
$$

Two times 22 is 44. Eighty-seven plus 44 is 131. Subtract 100 to get 31 because we have added 1 to our answer to make 2 instead of 1. Does this make sense? Yes, because we are finding the remainder from subtracting twice 378 from 787. The answer must be less than 100. Now bring down the next digit, 8.

```
400      12
378 | 45,678
 22     22ˣˣ
        787
         44
        318
```

Four hundred will not divide into 318. We now check with our actual divisor. Three hundred and seventy-eight will not divide into 318 so the next digit of the answer is zero and the 318 becomes the remainder.

```
400      120
378 | 45,678
 22     22ˣˣ
        787
         44
        318   remainder
```

The answer is 120 with 318 remainder (120 r318).

You have a choice

How would you solve this problem?

$1,410 \div 95 =$

You could solve this in two ways using this method. Let's see. Here is the first way.

```
100
95 | 1,410
 5
```

You could ask yourself, 'How many times does 100 divide into 1,400?' Fourteen times. Write down 14 as the answer.

Now for the remainder. Fourteen times 5 is 70. (Fourteen is 2 times 7. Multiply 5 by 2 to get 10, and then multiply 10 by 7 to get 70.) Seventy plus the remainder from 1,410 (10) makes 80.

```
100     14
95 | 1,410
 5      70
        80   remainder
```

The answer is 14 with 80 remainder (14 r80).

Here is the second way:

```
100     14
95 | 1,410
 5      5ˣ
       460
        20
        80   remainder
```

One hundred divides into 141 one time with 41 remainder.

One times 5 is 5, plus 41 remainder makes 46. Bring down the next digit, 0, to make 460.

One hundred divides into 460 four times with 60 remainder. Four times 5 is 20, plus the 60 remainder makes a remainder of 80.

The first method is easier, don't you think, and more suited to mental calculation? Try calculating the problem both ways in your head and see if you agree.

Possible complications

Here is an interesting example that shows some possible complications with using this method.

$3,456 \div 187 =$

We set out the problem:

```
200
187 | 3,456
(13)
```

We divide 200 into 300 of 345. The answer is 1, with 100 remainder.

We write it like this:

```
200      1
187 | 3¹,456
(13)
```

We write a small 1 above the hundreds digit (3 is the hundreds digit of 345) to signify the 100 remainder.

Now multiply 1 times 13, which equals 13. Add this to the 45 of 345, plus our 100 carried.

```
200      1
187 │ 3¹,456
 13     13
      158
```

What we are saying is this: if we have $345 in our pocket and purchase something for $187, we can hand over $200 to the seller and keep the other $145 in our pocket. We will be given $13 change which, added to the money in our pocket, makes $158 we have left.

We bring down the 6 for the final part of the calculation. Now we have 1,586 divided by 200.

Two hundred divides into 1,500 seven times with 100 left over. Don't forget to write this 100 remainder as a small 1 in the hundreds column. Because we have 7 times 13 remainder, plus the 186 from 1,586, we can see we can raise our answer by 1 to 8. Eight times 13 is easy. Eight times 10 is 80, plus 8 times 3 is 24, which adds up to 104.

```
200      18
187 │ 3¹,456
 13     13ˣ
      1,5¹86      –200
      ───────
        104
```

Because the extra 1 we have added to the answer accounts for another 200 we are dividing by, this must be subtracted from the remainder. I write '–200' to the side of my work to remind me.

$$\begin{array}{r} 200 \qquad 18 \\ \hline 187 \mid 3^1,456 \end{array}$$

200   18
187 | 3¹,456
(13)   13ˣ
       1,5¹86   −200
        104
         90

We add 186 to 104 to get 290. Now subtract the 200 written at the side to get our final remainder of 90. That is as complicated as it can get but you will find it becomes easy with practice. As long as you can keep track of what you are doing it won't be difficult. Practise some problems for yourself and you will find it becomes very easy.

Can we use this method to divide 34,567 by 937? Although 937 is not far from 1,000, we still have a larger difference — one that is not so easy to multiply.

Let's try it.

1000
937 | 34,567
(63)

The first calculation would be, 3,000 divided by 1,000. The answer is obviously 3. This is the first digit of the answer.

Now we have to multiply our circled number, 63, by 3.

Three times 60 is 180, and 3 times 3 is 9; the answer is 189. Write 189 below 3456 and add it to 456, for the remainder.

$$456 + 189 = 645$$

Now our calculation looks like this:

$$
\begin{array}{r|l}
1,000 \quad\; 3 \\
937 & 34,567 \\
\text{\small(63)} & \underline{189} \\
& 645
\end{array}
$$

Now bring down the next digit, 7.

$$
\begin{array}{r|l}
1,000 \quad\; 3 \\
937 & 34,567 \\
\text{\small(63)} & \underline{189}^{\text{x}} \\
& 6,457
\end{array}
$$

We now have to divide 6,457 by 1,000.

Six thousand divided by 1,000 is 6. Now we multiply 63 by 6. Is this difficult? No. Six times 60 is 360, plus $3 \times 6 = 18$, giving us 378.

Add this to 457 to get our remainder of 835.

$$
\begin{array}{r|l}
1,000 \quad\; 36 \\
937 & 34,567 \\
\text{\small(63)} & \underline{189}^{\text{x}} \\
& 6,457 \\
& \underline{378} \\
& 835 \quad \text{remainder}
\end{array}
$$

So, 34,567 divided by 937 is 36 with 835 remainder.

Test yourself

Try these problems for yourself. Calculate for a remainder, then to one decimal place.

(a)   456 ÷ 194 =   (b)   6,789 ÷ 288 =

(c)   5,678 ÷ 186 =   (d)   73,251 ÷ 978 =

How did you go? Here are the problems fully worked.

(a)   200
      194 | 456
      ⑥   12
              68   remainder

      200   2.35
      194 | 456.00
      ⑥   12ˣˣ
              680
               18
              980
               30
               10   remainder

(continued)

The answer is 2.4 to one decimal place.

(b)

```
 300        23
 288  |  6,789
 (12)      24ˣ
         1,029
           36
          165     remainder
```

```
 300        23.57
 288  |  6,789.00
 (12)      24ˣ ˣˣ
         1,0¹29
           36
         1,6¹50
           60
         2,100
           84
           84
```

The answer is 23.6 to one decimal place.

(c)

```
  200        30
  186 │ 5,678
 (14)    42ˣ
         098    remainder
```

$$
\begin{array}{r}
200 \quad 30 \\
186 \;\rvert\; 5{,}678 \\
(14) \quad 42^{x} \\
098 \quad \text{remainder}
\end{array}
$$

```
  200        30.52
  186 │ 5,678.00
 (14)    42ˣ ˣˣ
         0980
          70
         5¹00
          28
         128
```

The answer is 30.5 to one decimal place.

To multiply by 14 we simply multiply by 7 and double the answer (2 × 7 = 14).

(continued)

(d) 1,000   74

978 | 73,251

(22)   154^x

4,791

88

879   remainder

1,000   74.89

978 | 73,251.00

(22)   154^{x xx}

4,791

88

8,790

176

9,660

198

858

The answer is 74.9 to one decimal place.

In this problem we had to multiply by 22. This is easy if we remember that 22 is 2 times 11. It is easy to multiply by 2 and by 11 using our short cut. For instance, we had to multiply 8 times 22. Multiply 8 by 2, then multiple the answer by 11.

$8 \times 2 = 16$

$16 \times 11 = 176$

To divide by 19, 29 or 39, it would be easier to use our method of direct division, but when the divisor is just below 100, 200, a multiple of 100 or 1,000, you may find this method easier.

You should be able to solve problems like 1,312 divided by 96 in your head. You would divide by 100 minus 4. One hundred divides into 1,300 thirteen times, so you should be able to say immediately, 'Thirteen with 4 times 13 remainder, plus 12, or, 13 with 64 remainder.'

Then, if you want the answer to one decimal place, multiply 64 remainder by 10 and divide again. Six hundred and forty divided by 96 is 6 with 40 plus 24 remainder, which is 64. This is obviously going to repeat so you take the answer to as many decimal places as you like. The answer to three decimal places is 13.667.

Dividing by numbers just above 100

What do we do if the number we are dividing by is just above 100?

Imagine you are in a store and you need some cartons of long-life milk. A carton of milk costs $1.05. How many cartons can you buy if you have $5.45 in your pocket? You would immediately see you could buy 5 cartons and have some change.

Let's do the calculation.

We round off to a dollar, so $1.05 is 5 cents above a dollar. For each carton we buy, we are charged another 5 cents for each dollar we pay. So, if we buy 5 cartons, we pay 5 times 5 cents above the 5 dollars.

Let's try an example:

$537 \div 104 =$

```
       4     5
 104 │  537
        −20
```

17 remainder

One hundred divides five times into 500. But we are dividing by 104 so we have to take another 5 times 4 from the number we are dividing to get our remainder.

Comparing long division by adding to regular long division

To complete this chapter, let's compare long division by adding with the regular method for long division.

Example:

$705 \div 94 =$

The 'addition' method:

```
     100    7
  94 │  705
  (6)   42
        47
```

How many times does 100 divide into 705? Seven times.

Then we multiply 6 times 7 and add the answer to 5 (leftover from the 705) to get our remainder. It is easy to multiply 6 times 7 and easy to add 42 plus 5.

Now, let's compare this with regular long division.

$$
\begin{array}{r}
7 \\
94 \overline{\smash{)}\ 705} \\
\underline{658} \\
47
\end{array}
$$

How many times does 94 divide into 705? Seven times.

Then we multiply 7 times 94 to get our answer of 658. We subtract 658 from 705 to get our remainder.

Our method is much easier, don't you think?

Chapter 20
Checking answers (division)

Casting out nines is one of the most useful tools available for working with mathematics. I use it almost every day. Casting out nines is easy to use for addition and multiplication. Now we are going to look at how we use the method to check division calculations.

Changing to multiplication

When we looked at casting out nines to check a problem with subtraction, we found we often had to reverse the problem to addition. With division, we need to reverse the problem to one of multiplication. How do we do that?

Let's say we divide 24 by 6 to get an answer of 4. The reverse of that would be to multiply the answer by the divisor to get the original number we divided. That is, the reverse of $24 \div 6 = 4$ is $4 \times 6 = 24$.

That is not difficult.

To check the answer to the problem $578 \div 17 = 34$, we would use substitute numbers.

$$578 \div 17 = 34$$
$$20 \quad\quad 8 \quad\quad 7$$
$$2$$

Substituting, we have $2 \div 8 = 7$. That doesn't make sense. So, we do the calculation in reverse. We multiply: $7 \times 8 = 2$. Does 7 times 8 equal 2 with our substitute numbers?

$7 \times 8 = 56$

$5 + 6 = 11$

$1 + 1 = 2$

Seven times 8 does equal 2; our answer is correct.

Handling remainders

How would we handle $581 \div 17 = 34$, with 3 remainder? We would subtract the remainder from 581 to make the calculation correct without a remainder. We could either subtract the remainder first to get $578 \div 17 = 34$, which is the problem we already checked, or we do it by casting out nines, like this:

$$(581 - 3) \div 17 = 34$$

Reversed, this becomes $17 \times 34 = 581 - 3$.

Writing in the substitutes we get:

$$17 \times 34 = 581 - 3$$
$$8 \quad\quad 7 \quad\quad 5 \quad\quad 3$$

Or, $8 \times 7 = 5 - 3$.

$8 \times 7 = 56$

$5 + 6 = 2$

$5 - 3 = 2$

The answer checks correctly.

If you are not sure how to check a problem in division, try a simple problem like $14 \div 4 = 3$ r2.

Reverse the problem, subtracting the 2 remainder from 14 to get $3 \times 4 = 14 - 2$. Then apply the method to the problem that you want to check.

Finding the remainder with a calculator

When you carry out a division with your calculator, it gives your remainder as a decimal. Is there an easy way to find out the true remainder instead?

Yes, there is. If you divide 326 by 7 with a calculator, you get an answer of 46.571428 with an eight-digit calculator. What if you are trying to calculate items you have to divide up; how do you know how many will be left over?

The simple way is to subtract the whole number before the decimal and just get the decimal part of the answer. In this case you would just subtract 46, which gives 0.571428. Now multiply this number by the number you divided by. For our calculation, we multiply 0.571428 by 7 to get an answer of 3.999996. You round the answer off to 4 remainder.

Why didn't the calculator just say 4? Because it only works with numbers to a limited number of decimal places, so the answer is never exact.

Multiplying the decimal remainder will almost always give an answer that is fractionally below the correct remainder. You will be able to see this for yourself. If you are dividing 326 chairs into 7 classrooms at school, you know you won't have 46 in each room with 3.999996 chairs left over. You would have 4 chairs to keep for spares.

Bonus: casting twos, tens and fives

Just as it is possible to check a calculation by casting out the nines, you can cast out any number to make your check.

Casting out twos will only tell you if your answer should be odd or even. When you cast out twos, the only substitutes possible are 0 when the number is even and 1 when the number is odd. That is not very helpful.

When I was in primary school I sometimes checked answers by casting out the tens. All that did was check if the units digit of the answer was correct. Casting out the tens means you ignore every digit of a number except the units digit. Again, it is not very useful, but I did use it for multi-choice tests where a check of the units digit was sometimes enough to recognise the correct answer without doing the whole calculation.

Let's try an example. Which of these is correct?

$34 \times 72 =$

(a) 2,345　　(b) 2,448　　(c) 2,673　　(d) 2,254

Multiplying two even numbers can't give an odd number for an answer. (That is casting out twos.) That eliminates (a) and (c). To check the other two answers, we multiply the units digits of our problem together and get an answer of 8 (4 × 2 = 8). The answer must end with 8, so the answer must be (b).

Casting out fives is another option. For the above calculation, the substitute numbers would be exactly the same. But it can have its uses for checking multiplication by small numbers.

Let's try 7 × 8 = 56 as an example. Casting out fives we get 2 × 3 = 1.

We divide by 5 and just use the remainder. Again, we are only working with the units digits. Five divides once into 7 with 2 remainder; it divides once into 8 with 3 remainder and, just using the 6 of 56, we see there is 1 remainder.

Does it check? Let's see: 2 × 3 = 6 (which is the same as the units digit of 56), and 6 has a fives remainder of 1, just like our check. The problem with this is that 36, or even 41, would also have checked as correct using this method.

Let's try casting out sevens to check 9 × 8 = 72.

Nine and 8 have substitutes of 2 and 1, and 2 × 1 = 2. Two is our check answer.

We can cast out the 7 from 72 as it represents seven tens (7 × 10), which leaves us with 2. Our answer is correct. Casting out sevens was a better check than casting out twos, tens or fives because it involved all of the digits of the answer, but it is still too much trouble to be useful. Casting out nines makes far more sense, but it is still fun to experiment.

Casting out nines with minus substitute numbers

Can we use our method of casting out nines (substitute numbers) to check a simple calculation like 7 times 8? Is it possible to check numbers below 10 by casting out nines?

$$7 \times 8 = 56$$

The substitutes for 7 and 8 are 7 and 8, so it doesn't help us very much.

There is another way of casting out nines that you might like to play with. To check $7 \times 8 = 56$, you can subtract 7 and 8 from 9 to get minus substitute numbers.

Seven is 9 minus 2 and 8 is 9 minus 1, so our substitute numbers are −2 and −1. So long as they are both minus numbers, when you multiply them you get a plus answer, so you can treat them as simply 2 and 1.

Let's check our answer using the minus substitutes.

$$\begin{array}{ccccc} 7 & \times & 8 & = & 56 \\ 2 & & 1 & & 2 \end{array}$$

This checks out. Our substitute numbers, 2 and 1, multiplied give us the same answer as our substitute answer, 2.

If you weren't sure of the answer to 7×8 you could calculate the answer using the circles, or you can cast out the nines to double-check your answer.

You can have fun playing with this method. Let's check $2 \times 8 = 16$.

The minus substitutes are 7 and 1. Seven times 1 is 7. Seven is our check answer. The real answer adds to 7 (1 + 6 = 7), so our answer is correct.

Every number has two substitutes when you cast out the nines — a plus and a minus substitute. The number 25 has a positive substitute of 7 (2 + 5 = 7) and a negative substitute of −2 (9 − 7 = 2).

Let's try it with 8 × 8 = 64.

Eight is 9 minus 1, so −1 is our substitute for 8.

The substitute for 64 is 6 + 4 = 10; then, 1 + 0 = 1.

$$8 \quad \times \quad 8 \quad = \quad 64$$
$$1 \qquad 1 \qquad \quad 1$$

This only has a limited application but you can play and experiment. You could use it to check a subtraction where the substitute for the number you are subtracting is larger than the number you are subtracting from. This is because minus substitutes change a plus to a minus and a minus to a plus.

How does this help? Let's see.

If we want to check 12 − 8 = 4, our normal substitute numbers aren't much help.

$$12 \quad - \quad 8 \quad = \quad 4$$
$$3 \qquad 8 \qquad 4$$

Because using a minus substitute changes the sign — that is, a plus changes to minus and a minus changes to plus — −8 has a minus substitute of +1.

Our check now becomes 3 + 1 = 4, which we see is correct.

Of course, we wouldn't normally use this check for a problem like 12 − 8, but it is an option when we are working with larger numbers. For instance, let's check 265 − 143 = 122.

We find the substitutes by adding the digits:

$$265 \quad - \quad 143 \quad = \quad 122$$
$$13 \qquad\quad 8 \qquad\quad 5$$
$$4$$

Our substitute calculation becomes 4 − 8 = 5.

If we find the minus equivalent of −8 we get +1. So, our check becomes 4 + 1 = 5.

We can easily see now that our answer is correct.

Even though you might not use negative substitute numbers very often, they are still fun to play with, and, as you use them, you will become more familiar with positive and negative numbers. You also have another option for checking answers.

Chapter 21
Fractions made easy

When I was in primary school I noticed that many of my teachers had problems when they had to explain fractions. But fractions are easy.

Working with fractions

We work with fractions all the time. If you tell the time, chances are you are using fractions (half past, a quarter to, a quarter past, etc.). When you eat half an apple, this is a fraction. When you talk about football or basketball (half-time, second half, three-quarter time, etc.), you are talking about fractions.

We even add and subtract fractions, often without realising it or thinking about it. We know that two quarters make a half. Half-time during a game is at the end of the second quarter. Two quarters make one half.

If you know that half of 10 is 5, you have done a calculation involving fractions. You have half of your 10 fingers on each hand.

So what is a fraction?

A fraction is a number such as ½. The top number — in this case 1 — is called the *numerator*, and the bottom number — in this case 2 — is called the *denominator*.

The bottom number, the denominator, tells you how many parts something is divided into. A football game is divided into four parts, or quarters.

The top number, the numerator, tells you how many of these parts we are working with — three-quarters of a cake, or of a game.

Writing ½ is another way of saying 1 divided by 2. Any division problem can be expressed as a fraction: ⁶⁄₃ means 6 divided by 3; ¹⁵⁄₅ means 15 divided by 5. Even 5,217 divided by 61 can be written as $^{5,217}/_{61}$. We can say that any fraction is a problem of division. Two-thirds (⅔) means 2 divided by 3.

We often have to add, subtract, multiply or divide parts of things. That is another way of saying we often have to add, subtract, multiply or divide fractions.

I am going to give you some hints to make your work with fractions easier. (If you want to read and learn more about fractions you can read my book *Speed Mathematics*.)

Here is how we add and subtract fractions.

Adding fractions

If we are adding quarters the calculation is easy. One-quarter plus one-quarter makes two-quarters, or a half. If you add another quarter you have three-quarters. If the denominators are the same, you simply add the numerators. For instance, if you wanted to add one-eighth plus two-eighths, you would have an answer of three-eighths. Three-eighths plus three-eighths gives an answer of six-eighths.

How would you add one-quarter plus one-eighth?

$$\frac{1}{4} + \frac{1}{8} =$$

If you change the quarter to $\frac{2}{8}$ then you have an easy calculation of $\frac{2}{8} + \frac{1}{8}$.

It is not difficult to add $\frac{1}{3}$ and $\frac{1}{6}$. If you can see that $\frac{1}{3}$ is the same as $\frac{2}{6}$ then you are just adding sixths together. So $\frac{2}{6}$ plus $\frac{1}{6}$ equals $\frac{3}{6}$. You just add the numerators.

This can be easily seen if you are dividing slices of a cake. If the cake is divided into 6 slices, and you eat 1 piece ($\frac{1}{6}$) and your friend has 2 pieces ($\frac{2}{6}$), you have eaten $\frac{3}{6}$ of the cake. Because 3 is half of 6 you can see that you have eaten half of the cake.

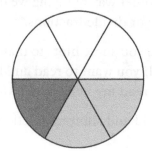

$\frac{1}{6} + \frac{2}{6} = \frac{1}{2}$

Adding fractions is easy. Here is how we would add ⅓ plus ⅖. The standard method is to change thirds and fifths into the same parts like we did with thirds and sixths.

Here is an easy way to solve ⅓ plus ⅖ that you probably won't be taught in school. Firstly, we multiply crossways and add the answers to get the numerator of the answer.

$$\frac{1}{3} \diagdown + \diagup \frac{2}{5} \quad =$$

$1 \times 5 = 5$

$3 \times 2 = 6$

$5 + 6 = 11$

Eleven is the top number (the numerator) of the answer. Now we multiply the bottom numbers (denominators) to find the denominator of the answer.

$3 \times 5 = 15$

The answer is ¹¹⁄₁₅.

$$\frac{1}{3} \diagup\diagdown \frac{2}{5} \quad = \quad \frac{5+6}{15} \quad = \quad \frac{11}{15}$$

Easy.

Here is another example:

$$\frac{3}{8} + \frac{1}{7} =$$

Multiply crossways.

$3 \times 7 = 21$

$8 \times 1 = 8$

$$\frac{21 + 8}{8 \times 7}$$

We add the totals for the numerator, which gives us 29. Then we multiply the denominators:

$8 \times 7 = 56$

This is the denominator of the answer. Our answer is $^{29}\!/_{56}$.

Let's try one more.

$$\frac{2}{5} \diagdown\!\!\!\!+\!\!\!\!\diagup \frac{3}{8} = \frac{16 + 15}{40} = \frac{31}{40}$$

Simplifying answers

In each of these examples, the answer we got was the final answer.

Let's try another example to see one more step we need to take to complete the calculation.

$$\frac{1}{3} + \frac{2}{9} = \frac{9 + 6}{27} = \frac{15}{27}$$

We have the correct answer, but can the answer be simplified?

If both the numerator and denominator are even, we can divide them by 2 to get a simpler answer. For example, $^4\!/_8$ could be simplified to $\frac{1}{2}$ because both 4 and 8 are divisible by 4.

We see with the above answer of $^{15}\!/_{27}$ that 15 and 27 can't be divided by 2, but they are both evenly divisible by 3 ($15 \div 3 = 5$ and $27 \div 3 = 9$).

The answer of $^{15}\!/_{27}$ can be simplified to $^5\!/_9$.

Each time you calculate fractions, you should give the answer in its simplest form. Check to see if the numerator and denominator are both divisible by 2, 3 or 5, or any other number. If so, divide them to give the answer in its simplest form. For instance, $^{21}/_{28}$ would become $^3/_4$ (21 ÷ 7 = 3 and 28 ÷ 7 = 4).

Let's try another example, $^2/_3 + ^3/_4$.

$$\frac{2}{3} + \frac{3}{4} = \frac{8+9}{12} = \frac{17}{12}$$

In this example the top number (numerator) is larger than the bottom number (denominator). When this is the case, we have to divide the top number by the bottom, and express the remainder as a fraction. So, 12 divides one time into 17 with 5 remainder. Our answer is written like this:

$1^{5/12}$

Test yourself

Try these for yourself:

(a) $^1/_4 + ^1/_3 =$ (b) $^2/_5 + ^1/_4 =$

(c) $^3/_4 + ^1/_5 =$ (d) $^1/_4 + ^3/_5 =$

How did you go? The answers are:

(a) $^7/_{12}$ (b) $^{13}/_{20}$ (c) $^{19}/_{20}$ (d) $^{17}/_{20}$

If your maths teacher says you must use the lowest common denominator to add fractions, is there an easy way to find it? Yes, you simply multiply the denominators and then see if the answer can be simplified. If you multiply the denominators you certainly have a common denominator, even if it isn't the lowest. (You should understand the standard method of adding and subtracting fractions as well as the methods taught in this book.)

A short cut

Here is a short cut. If the numerators are both 1, we add the denominators to find the numerator of the answer (top number), and we multiply the denominators to find the denominator of the answer. These can easily be done in your head.

Here is an example:

$$\frac{1}{4}+\frac{1}{5}=\frac{4+5}{4\times5}=\frac{9}{20}$$

For you just to look at a problem like this and call out the answer will make everyone think you are a genius!

Here's another: if you want to add ⅓ plus ⅐ you add 3 and 7 to get 10 for the numerator, then you multiply 3 and 7 to get 21 for the denominator.

$$\frac{1}{3}+\frac{1}{7}=\frac{3+7}{3\times7}=\frac{10}{21}$$

Test yourself

Try these for yourself. Do them all in your head.

(a) $\frac{1}{4} + \frac{1}{3} =$

(b) $\frac{1}{5} + \frac{1}{4} =$

(c) $\frac{1}{3} + \frac{1}{5} =$

(d) $\frac{1}{4} + \frac{1}{7} =$

How did you go? The answers are:

(a) $\frac{7}{12}$ (b) $\frac{9}{20}$ (c) $\frac{8}{15}$ (d) $\frac{11}{28}$

You should have had no trouble with those.

Subtracting fractions

A similar method is used for subtraction. You multiply the numerator of the first fraction by the denominator of the second fraction. You then multiply the denominator of the first fraction by the numerator of the second, and subtract the answer. You find the denominator of the answer in the usual way; by multiplying the denominators together.

Here's an example:

$$\frac{2}{3} \diagtimes \frac{1}{4} \ = \ \frac{8-3}{12} \ = \ \frac{5}{12}$$

What we did was multiply $2 \times 4 = 8$, and then subtract $3 \times 1 = 3$, to get 5. Five is the numerator of our answer. We multiplied $3 \times 4 = 12$ to get the denominator. You could do that entirely in your head.

Test yourself

Try these for yourself. Try them without writing anything down except the answers.

(a) $\frac{1}{2} - \frac{1}{7} =$ (b) $\frac{1}{3} - \frac{1}{5} =$

(c) $\frac{1}{5} - \frac{1}{7} =$ (d) $\frac{1}{2} - \frac{1}{3} =$

The answers are:

(a) $\frac{5}{14}$ (b) $\frac{2}{15}$ (c) $\frac{2}{35}$ (d) $\frac{1}{6}$

If you made any mistakes, read the section through again.

Try some for yourself and explain this method to your closest friend. Tell your friend not to tell the rest of your class.

A short cut

Just as with addition, there is an easy short cut when you subtract fractions when both numerators are 1. The only thing to remember is to subtract backwards.

$$\frac{1}{2} - \frac{1}{5} = \frac{5-2}{10} = \frac{3}{10}$$

Again, these are easily done in your head. If the other kids in your class have problems with fractions they will be really impressed when you just look at a problem and call out the answer.

Test yourself

Try these for yourself (do them in your head):

(a) $\frac{1}{4} - \frac{1}{5} =$

(b) $\frac{1}{2} - \frac{1}{6} =$

(c) $\frac{1}{4} - \frac{1}{7} =$

(d) $\frac{1}{3} - \frac{1}{7} =$

The answers are:

(a) $\frac{1}{20}$

(b) $\frac{1}{3}$

(c) $\frac{3}{28}$

(d) $\frac{4}{21}$

Can you believe that adding and subtracting fractions could be so easy?

Multiplying fractions

What answer would you expect if you had to multiply one-half by two? You would have two halves. How about multiplying a third by three? You would have three thirds. How much is two halves? How much is three thirds?

If you split something into halves and bring the two halves together, how much have you got? A whole. You could say it mathematically like this: $\frac{1}{2} \times 2 = 1$.

If you cut a pie into thirds, and you have all three thirds, you have the whole pie. You could say it mathematically like this: $\frac{1}{3} \times 3 = 1$.

Here is another question: what is half of 12? The answer is obviously 6. But, what did you do to find your answer? To find half of a number you divide by 2. Or, you could say, you multiplied by one-half. If you can understand that, you can multiply fractions.

Here is how we do it. We simply multiply the numerators to get the numerator of the answer, and we multiply the denominators to get the denominator of the answer. Easy!

Let's try it to find half of 12.

$$\frac{12}{1} \times \frac{1}{2} = \frac{12}{2} = \frac{6}{1} = 6$$

Any whole number can be expressed as that number over (divided by) 1. So 12 is the same as $^{12}\!/_1$.

Let's try another:

$$\frac{2}{3} \times \frac{4}{5} =$$

To calculate the answer, we multiply 2 × 4 to get 8. That is the top number of the answer. To get the bottom number of our answer we multiply the bottom numbers of the fractions.

$$3 \times 5 = 15$$

The answer is $^{8}\!/_{15}$. It is as easy as that. What is half of 17?

$$\frac{17}{1} \times \frac{1}{2} =$$

Multiply the numerators.

$$17 \times 1 = 17$$

Seventeen is the numerator of the answer.

Multiply the denominators.

$1 \times 2 = 2$

Two is the denominator of the answer.

So, the answer is 17 divided by 2, which is 8 with 1 remainder. The remainder goes to the top (numerator), and the 2 at the bottom remains where it is for the answer. Our answer is 8½.

Dividing fractions

In the previous section we multiplied 17 by ½ to find half of 17. How would you divide 17 by ½?

Let's assume we have 17 oranges to divide among 30 children.

If we cut each orange in half (divide by ½), we would have enough for everyone, plus some left over.

If we cut an orange in half we get 2 pieces for each orange, so we had 17 oranges but 34 orange halves. Dividing the oranges makes the number bigger. So, to divide by one-half we can actually multiply by 2, because we get 2 halves for each orange.

So $^{17}\!/_1 \div$ ½ is the same as $^{17}\!/_1 \times$ $^2\!/_1$.

To divide by a fraction, we turn the fraction we are dividing by upside down and make it a multiplication. That's not too hard!

Who said fractions are difficult?

Test yourself

Try these problems in your head:

(a) $\frac{1}{2} \div \frac{1}{2} =$ (b) $\frac{1}{3} \div \frac{1}{2} =$

(c) $\frac{2}{5} \div \frac{1}{4} =$ (d) $\frac{1}{4} \div \frac{1}{4} =$

Here are the answers. How did you go?

(a) 1   (b) $\frac{2}{3}$ (c) $\frac{8}{5}$ or $1\frac{3}{5}$ (d) 1

Changing vulgar fractions to decimals

The fractions we have been dealing with are called vulgar fractions, or common fractions. To change a vulgar fraction to a decimal is easy; you simply divide the numerator by the denominator (the top number is divided by the bottom number).

To change ½ to a decimal you divide 1 by 2. Two won't divide into 1, so the answer is 0 and we carry the 1 to make 10. The decimal in the answer goes below the decimal in the number we are dividing. Two divides into 10 exactly 5 times.

The calculation looks like this:

```
2 | 1.000
     0.5
```

Here's another: one-eighth (⅛) is 1 divided by 8. One divided by 8 equals 0.125.

That is all there is to it. Try a few more yourself.

Chapter 22
Direct multiplication

Everywhere I teach my methods I am asked, how would you multiply these numbers? Usually I will show people how to use the methods you have learnt in this book, and the calculation is quite simple. There are often several ways to use my methods, and I delight in showing different ways to make the calculation simple.

Occasionally I am given numbers that do not lend themselves to my methods with a reference number and circles. When this happens I tell people that I use direct multiplication. This is traditional multiplication, with a difference.

Multiplication with a difference

For instance, if I were asked to multiply 6 times 17, I wouldn't use my method with the circles as I think it is not the easiest way to solve this particular problem. I would simply multiply 6 times 10 and add 6 times 7.

$6 \times 10 = 60$
$6 \times 7 = 42$
$60 + 42 = 102$ *Answer*

How about 6 times 27?

Six times 20 is 120 (6 × 2 × 10 = 120). Six times 7 is 42. Then, 120 + 42 = 162. The addition is easy: 120 plus 40 is 160, plus 2 is 162.

This is much easier than working with positive and negative numbers.

It is easy to multiply a two-digit number by a one-digit number. For these types of problems, you have the option of using 60, 70 and 80 as reference numbers. This means that there is no gap in the numbers up to 100 that are easy to multiply.

Let's try a few more for practice:

$7 \times 63 =$

You could use two reference numbers for this, so we will try both methods.

Firstly, let's use direct multiplication.

$7 \times 60 = 420$
$7 \times 3 = 21$
$420 + 21 = 441$

That wasn't too hard.

Now let's use 10 and 70 as reference numbers.

(10 × 7)   7   ×   63   =

⊖(-3)   ⊖(-7)

⊖(-21)

63 − 21 = 42
42 × 10 = 420
3 × 7 = 21
420 + 21 = 441

(10 × 7)   7   ×   63   420

⊖(-3)   ⊖(-7)   +   21

⊖(-21)   441   *Answer*

The calculations were almost identical in this case, but I think the direct method was easier.

How about 6 × 84?

6 × 80 = 480
6 × 4 = 24
480 + 24 = 504

Now using two reference numbers we have:

(10 × 9)   6   ×   84   =

⊖(-4)   ⊖(-6)

⊖(-36)

Subtracting 36 from 84 we get 480. Four times 6 is 24, then 480 + 24 = 504.

Happily, we get the same answer. This time the direct method was definitely easier, although, again, the calculations ended up being very similar.

Let's try one more: 8 × 27.

By direct multiplication we have:

$8 \times 20 = 160$
$8 \times 7 = 56$
$160 + 56 = 216$

Now using two reference numbers we have:

$$(10 \times 3) \qquad 8 \quad \times \quad 27 \quad = \quad 210$$

$$\boxed{-2} \qquad \boxed{-3} \qquad + \underline{6}$$

$$\boxed{-6} \qquad\qquad\qquad 216 \quad \textit{Answer}$$

This time the calculation was easier using two reference numbers.

Is it difficult to use direct multiplication when we have a reference number of 60, 70 or 80? Let's try it for 67 times 67. We will use 70 as our reference number.

$$\boxed{70} \qquad 67 \quad \times \quad 67 \quad =$$

$$\boxed{-3} \qquad\quad \boxed{-3}$$

We subtract 3 from 67 to get 64. Then we multiply 64 by our reference number, 70. Seventy is 7 times 10, so we multiply by 7 and then by 10.

Seven times 60 is 420 ($6 \times 7 = 42$, then by 10 is 420).

Seven times 4 is 28. Then, $420 + 28 = 448$, so 7 times 64 is 448, and 70 times 64 is 4,480.

Multiply the numbers in the circles: $3 \times 3 = 9$. Then:

$4,480 + 9 = 4,489$ *Answer*

Even if you go back to using your calculator for such problems, you have at least proven to yourself that you can do them yourself. You have acquired a new mathematical skill.

How would you use this method to solve 34 × 76?

One method would be to use 30 as a reference number and use factors of 2 × 38 to replace 76. You could also solve it directly using direct multiplication. Here is how we do it:

$$\begin{array}{r} 76 \\ \times\underline{34} \end{array}$$

We will multiply from right to left (as is most common), and then try it again from left to right.

Firstly, we begin with the units digits:

$6 \times 4 = 24$

We write the 4 and carry the 2. Then we multiply crossways and add the answer.

$7 \times 4 = 28$
$3 \times 6 = 18$
$28 + 18 = 46$

(To add 18 you could add 20 and subtract 2.) Add the carried 2 to get 48. Write 8 and carry 4.

Now multiply the tens digits:

$3 \times 7 = 21$

Add the 4 carried to get 25. Now we write 25 to finish the answer. The calculation was really done using traditional methods but written in one line. The finished calculation looks like this:

$$\begin{array}{r} 76 \\ \times\ \underline{34} \\ 25^48^24 \end{array}$$

Here are the steps for this problem:

$6 \times 4 = 24$

$$\begin{array}{r} 76 \\ | \\ \times\ \underline{34} \\ ^24 \end{array}$$

$7 \times 4 = 28$
$3 \times 6 = 18$
$28 + 18 + 2 \text{ carried} = 48$

$$\begin{array}{r} 76 \\ \times \\ \times\ \underline{34} \\ ^48^24 \end{array}$$

$7 \times 3 = 21 + 4 \text{ carried} = 25$

$$\begin{array}{r} 76 \\ | \\ \times\ \underline{34} \\ 25^48^24 \end{array}$$

Each digit is multiplied by every other digit. The full working looks like this:

$$\begin{array}{r} 76 \\ \times\ \underline{34} \\ 25^48^24 \end{array}$$

Solving from right to left is probably easiest if you are using pen and paper. If you want to solve the problem mentally then you can go from left to right. Let's try it.

Thirty times 70 is $3 \times 7 \times 100$. The 100 represents the two zeros from 30 and 70.

$3 \times 7 = 21$
$21 \times 100 = 2,100$

I would say to myself, twenty-one hundred.

Now we multiply crossways: 3×6 and 7×4, and then multiply the answer by 10.

$3 \times 6 = 18$
$7 \times 4 = 28$
$18 + 28 = 46$
$46 \times 10 = 460$

Our subtotal was 2,100. To this we add 400, then we add 60.

$2,100 + 400 = 2,500$
$2,500 + 60 = 2,560$

We are nearly there. Now we multiply the units digits, and add.

$4 \times 6 = 24$
$2,560 + 24 = 2,584$

Try doing the problem yourself in your head. You will find it is easier than you think. Calculating from left to right means there are no numbers to carry.

The calculation is not as difficult as it appears, but your friends will be very impressed. You just need to practice shrugging your shoulders and saying, 'Oh, it was nothing.'

If you can't find an easy way to solve a problem using a reference number then direct multiplication might be your best option.

Direct multiplication using negative numbers

I debated with myself whether this section should be included in the book. If you find it difficult, don't worry about it. Try it anyway. But you may find this quite easy. It involves using positive and negative numbers to solve problems in direct multiplication. If you try it and you think it is too difficult, forget it — or come back to it in a year's time and try it again. I enjoy playing with this method. See what you think. Here is how it works.

If you are multiplying a number by 79 it may be easier to use 80 – 1 as your multiplier. Multiplying by 79 means you are multiplying by two high numbers and you are likely to have high subtotals. Multiplication by 80 – 1 might be easier. Using 80 – 1, 8 becomes the tens digit and −1 is the units digit. You need to be confident with negative numbers to try this.

Let's have a go:

$68 \times 79 =$

We set it out as:

```
6   8
8  −1
```

We begin by multiplying the units digits. Eight times minus 1 is minus 8. We don't write −8; we borrow 10 from the tens column and write 2, which is left over when we minus the 8. We carry −1 (ten), which we borrowed, to the tens column.

The work looks like this:

$$
\begin{array}{cc}
6 & 8 \\
8 & -1 \\
\hline
 & {}^{-1}2
\end{array}
$$

Now we multiply crossways. Eight times 8 is 64, and 6 times −1 is −6.

$64 - 6 = 58$

We subtract the 1 that was carried (because it was −1) to get 57. We write the 7 and carry the 5.

$$
\begin{array}{cc}
6 & 8 \\
8 & -1 \\
\hline
{}^{5}7 & {}^{-1}2
\end{array}
$$

For the final step we multiply the tens digits.

$6 \times 8 = 48$

We add the 5 that was carried.

$48 + 5 = 53$

The answer is 5,372.

$$
\begin{array}{cc}
6 & 8 \\
8 & -1 \\
\hline
53^{5}7 & {}^{-1}2
\end{array}
$$

Test yourself

Try these for yourself:

(a) 64 × 69 =

(b) 34 × 78 =

Here are the answers:

(a) 4,416   (b) 2,652

Let's work through the second problem together.

$$
\begin{array}{r}
3 \quad 4 \\
8 \; {-}2 \\
\hline
26^25 \; {}^{-1}2
\end{array}
$$

We begin by multiplying 4 times −2, which equals −8. Borrowing 10 we have 2 left. We carry the −10 to the next column as −1 (times 10).

Now we multiply crossways. Eight times 4 is 32, plus 3 times −2 is −6. Adding −6 is the same as subtracting 6 from 32, which gives us 26. Don't forget the carry of −1, so we end up with 25. Write the 5 and carry the 2. For the final step we multiply the tens digits: 3 × 8 = 24. We add the 2 which we carried to get 26. We now have our answer, 2,652.

This is simply a method you might like to play with. Experiment with problems of your own.

Chapter 23
Putting it all into practice

Often when people read books on high-speed maths and mathematical short cuts, they ask, 'How do I remember all of that?' They are overwhelmed by the amount of information and they simply say, I will never remember all of that stuff. I might as well forget about it.

Is there a possibility of this happening with the information in this book? There is always a possibility, but it is not likely. Why? Because books of mathematical short cuts that are easily forgotten are just that — a series of unconnected short cuts that have to be memorised for special occasions. And when the special occasion occurs, we either forget to use the short cut or we can't remember how to use it.

This book is different because it teaches a philosophy for working with mathematics. It teaches broad strategies that become part of the way we think. Putting the methods you have learnt in this book into practice will affect the way you think and calculate in almost every instance.

For instance, our methods of division apply to all division problems — they are not short cuts applicable to some isolated cases. You can apply the method of checking answers (casting out nines) to almost any calculation in arithmetic. It should become an automatic part of any calculation. We add and subtract just about every day. The methods in this book apply every day. If you are using the methods every day, how can you forget them?

This book doesn't teach short cuts — it teaches basic strategies. Practise the strategies and you will find you are automatically using factors for simple multiplication and division. You will use the methods for addition and subtraction. My philosophy has always been, if there is a simple way to do something, why do it the hard way?

Some of my students have even asked me if using my learning methods and maths methods is really cheating. We have an unfair advantage over the other kids. Is it ethical? I tell them that the better students have better methods anyway — that is what makes them better students. They say, but the other students could do as well as me if they learnt my methods. Sure they could. If you feel strongly about it, teach the other kids yourself, or lend them your book.

Most of the methods taught are 'invisible'. That is, the only difference is what goes on inside your head. If someone looked over your shoulder while you performed a subtraction in class, they wouldn't see by watching your work that you are doing anything different to the rest of the class. The same goes for long division, unless you are using the direct long-division method. So, every time you perform a calculation you have the choice to use the old method or use the easy method.

You often hear people say they have attended a class or seminar, the methods were great and easy, but they haven't used any of it since. They have wasted their money. That is why I tell my students to make my methods part of their life. You don't have to perform exercises every day and practise drills. That is a sure way to give up. Just use the methods every day as you have the need and opportunity. If you have an opportunity to use one of my methods and forget how it works, don't worry about it. Just read the chapter again, practise the methods by solving the examples as you read them, and make sure you are ready next time.

Also, it doesn't hurt to show off what you can do. Tell your friends you know your tables up to the 20 times table or the 25 times table. Then have them call out any combination for you to solve. They can check your answer with a calculator. Even if you feel you are slow giving the answer, they will just think this is one combination you don't know very well. I think most of your friends would prefer to think you have memorised the tables than to think you can calculate the answers in a flash. Then, when you are finished, 'confess' to your friends that you don't have them memorised, you actually worked them out as they gave them to you. They won't know whether to believe you or not! It is a fun way to practise and to learn your tables — and a fun way to show off your skills.

Advice for geniuses

I have often said that people think you are very intelligent if you are very good at mathematics. People will treat you differently; your friends, your classmates and your teachers. Everyone will think you are extra smart. Here is some advice for handling your new status.

Firstly, don't explain how you do everything. Author Isaac Asimov tells how he explained how he solved a mathematical problem to someone serving him in a store. When she heard the explanation she said scornfully, 'Oh, it was just a trick.' He hadn't used the method she had been taught at school. It was almost like he had cheated.

My methods are not tricks. Sometimes people will introduce me as someone who teaches maths tricks. I never like that. 'Tricks' to me implies trickery — you aren't really doing what you appear to be.

Also, people react with the comment, 'Oh, for a moment I thought you had done something clever.' They also often add, 'Well, anyone could do that,' once they know how you did it. When I am giving a class I will usually say, 'That's what makes my methods so good. Anyone *can* do it!'

Appendix A:
Using the methods in the classroom

Children often ask me, 'How do I use the methods in the classroom? Won't my teacher object if I use different methods?'

It is a fair question. Here is the answer I give.

Firstly, many of the methods taught in this book are 'invisible'. That is, the difference is what you say in your head. With addition and subtraction problems the layout and written calculations are the same; what is different is what you say to yourself in your head. When you are subtracting 8 from 5 and you borrow 10 to make it 8 from 15; you say to yourself 8 from 10 is 2, plus 5 is 7. You write down the 7. The students who subtract 8 from 15 write down 7 as well (so long as they don't make a mistake), so anyone looking over your shoulder would have no idea you are using a different method.

The same goes for subtracting 3,571 from 10,000. You set the problem out the same as everyone else, but again, what you say inside your head is different. You subtract each digit from 9 and the final digit from 10. No-one looking at your finished calculation would know you did anything different.

If you were asked to subtract 378 from 613 you would again write the problem down as usual. Then you could find the answer in your head by subtracting 400 and adding 22.

$$613 - 400 = 213$$
$$213 + 22 = 235$$

Using this method there is no carrying or borrowing. You just write the answer and then check by casting nines.

The same goes for all mental calculation; that is, all calculations where you write nothing down. No-one knows what method you use. Students and teachers have always used different methods, even for such simple problems as 56 minus 9, as I pointed out earlier in the book.

Also, the higher the grade you are in, the less of a problem this will be.

Now let's look at where it can make a difference. Let's say you are asked to multiply 355 by 52. You set out the problem as the teacher requires.

$$
\begin{array}{r}
355 \\
\times \quad 52 \\
\hline
710 \\
17,750 \\
\hline
18,460 \\
\end{array}
$$

Here is the finished calculation. Can you now make use of what you have learnt in this book? Certainly you can. You can cast out nines to check your answer.

$$355 \times 52 = \cancel{18},460$$
$$13 \qquad 7 \qquad 10$$
$$4 \qquad \qquad 1$$

The substitutes are $4 \times 7 = 1$, which we can see is correct because 4 times 7 equals 28, and 2 plus 8 is 10, which adds to 1.

Had you found you had made a mistake, you could cast out nines to find where you made it. You would find the substitute number for each part of the calculation. It would look like this.

$$
\begin{array}{rrr}
355 & & 4 \\
\times \quad 52 & & 7 \\
\hline
710 & & 8 \\
17{,}750 & 20 & 2 \\
\hline
\cancel{18}{,}460 & 10 & 1 \\
\end{array}
$$

To check, you would use the substitutes like this. Firstly you multiplied 355 by 2 to get 710. Now you multiply the substitute.

$$4 \times 2 = 8$$

Our substitute answer is 8, so this part of the calculation is correct.

Next, we multiplied 355 by 5 (or 50) to get 17,750. Multiply the substitutes again.

$$5 \times 4 = 20$$

The substitute answer is the same, so you can accept this part of the calculation as correct.

Next you added 710 and 17,750 to get your final answer of 18,460. We add the substitutes of 8 + 2 = 10, and this checks with your calculation. In this case your answer is correct but, if you had made a mistake, you not only would have found it, but you would also know where you made it. You would know if the mistake was in the multiplication, and which part, or if the mistake was in the addition.

I would make these checks with pencil and then erase them.

In practice, I think any teacher would be intrigued by your method of checking. I have never heard of these methods causing problems for students in the classroom.

Most importantly, once you earn a reputation for being mathematically gifted, no-one will worry about your methods — they will expect you to be different.

Appendix B:
Working through a problem

What you say inside your head while you perform mental mathematical calculations is very important. You can double the time it takes to solve a problem by saying the wrong things, and halve the time by saying the right things. That is why I tell you what to say, for instance, when you are multiplying by 11 in your head. To multiply 24 by 11, you would automatically see that 2 plus 4 is less than 10, so you would look at the 2 and immediately start saying, 'Two hundred and ...'. While you are saying that you would see that 2 plus 4 means you should say, '... sixty- ...', and in the same breath you would say, '... four.' So you would just say, 'Two hundred and sixty-four.'

To multiply 14 by 15 you would simply add 5 to 14 to get 19. You say in your head, 'One hundred and ninety ...'. Four times 5 is 20. Adding 20 gives, 'Two hundred and ten.' You would say 'One-ninety, two-ten.' The less you say, the faster you will be. Practise by yourself.

You don't need to recite the whole procedure. You wouldn't say, 'Fourteen plus five is one hundred and ninety. Now we multiply four times five. Four times five is twenty. One hundred and ninety plus twenty is two hundred and ten.' That would take you much longer, and seems like a whole lot more work. Just say the subtotals and totals or, if you can, just the answer.

When you practise any of the methods — especially if you want to show off in front of your friends or class — try to anticipate what is coming. If you see you will have to carry, say the next number with the carried number already added. For instance, let's multiply 11 times 84. You would immediately see that 8 plus 4 is 12 or, at least, more than 10. So you would know to carry 1 and start calling out, 'Nine hundred and …' You provide the 2 from 12 to give '… twenty- …', and then say the 4 at the end, giving 924. You would call the answer in one breath: 'Nine hundred and twenty-four.'

Often it helps to call carry numbers what they are. That is, if you are carrying 3 and the 3 represents 300, say three hundred. If it represents 30, say thirty. This can be useful in multiplication and addition.

Usually, the biggest hurdle to calling immediate answers is the feeling that the calculation is 'too hard'. Practise for yourself and you will see how easy it is to give immediate answers off the top of your head. This applies if you are saying the answer out loud or just calculating in your head. Practise some of the strategies with your closest friend — someone who won't embarrass you if you make a mistake.

Practise the problems in this book in your head. The first time may seem difficult, the second time is easier, and by the fifth time you will wonder why you ever thought they were hard. You will also find that this will build your concentration skills.

Appendix C:
Learn the 13, 14 and 15 times tables

When you know your multiplication tables up to the 5 times table you can multiply most problems in your head. If you know your 2, 3, 4 and 5 times tables up to 10 times each number, then you also know part of your 6, 7, 8 and 9 times tables. For the 6 times table you know 2 times 6, 3 times 6, 4 times 6, and 5 times 6. The same applies to the 7, 8 and 9 times tables. As you learn your higher tables, mental calculations become much easier. It is easy to learn your 11 times table, and the 12 times table is not difficult. Six times 12 is 6 times 10 plus 6 times 2. Six times 10 is 60 and 6 times 2 is 12. It is easy to add the answers for 6 times 12 to get 72.

The way tables used to be taught was difficult and boring. We have seen how easy it is to learn your tables and how it can be done much more quickly than in the olden days.

It is usual to learn tables up to the 12 times table. It is not hard to learn the 13 times table if you know the 12 times table. If you know 12 threes are 36, just add another 3 to get 39. Twelve threes plus one more 3 makes 13 threes. If you know 12 fours are 48, just add another 4 to get 52.

Then, when you know your 13 times table, learn the 14 times table. You have two ways to do this. You can factorise 14 to 2 times 7. Then you just double the number and multiply by 7, or multiply by 7 and double the answer. So, 6 times 14 would be 6 times 7 (42), then doubled is 84. That was easy. Also you could have doubled the 6 first to get 12, and multiply 12 by 7 to get 84 again. If you know your 12 times table, that is easy as well. Another alternative is to say 6 times 10 is 60, plus 6 times 4 is 24. Add 60 and 24 for the answer, 84. Calculate these often and you will learn your tables without even trying. Then you will find you can multiply and divide directly by 13, 14 and 15. This will give you an advantage over the other kids.

Appendix D:

Tests for divisibility

It is often useful to know whether a number is evenly divisible by another number. Here are some ways you can find out without doing a full division.

Divisibility by 2: All even numbers are divisible by 2; that is, all whole numbers for which the final digit is 0, 2, 4, 6 or 8.

Divisibility by 3: If the sum of the digits is divisible by 3, then the number is also divisible by 3. For instance, the digit sum of 45 is 9 (4 + 5 = 9), which is divisible by 3, so 45 is divisible by 3.

Divisibility by 4: If the last two digits are divisible by 4, the number is divisible by 4. This is because 100 is evenly divisible by 4 (100 = 4 × 25).

There is an easy short cut. Don't worry how long the number is; if the tens digit is even, forget it and check if the units digit is divisible by 4. (It must be either 4 or 8.) If the tens digit is odd, carry 1 to the units digit and check if it is divisible by 4. That is because 20 is evenly divisible by 4 (4 × 5 = 20).

For instance, let's check if 15,476 is divisible by four. The tens digit, 7, is odd, so we carry 1 to the units digit, 6, to make 16. Sixteen is evenly divisible by 4, so 15,476 is evenly divisible by 4.

Is 593,768 divisible by 4? The tens digit, 6, is even so we can ignore it. The units digit, 8, is evenly divisible by 4, so 593,768 is evenly divisible by 4.

Divisibility by 5: Five is easy. If the last digit of the number is 5 or 0, then the number is evenly divisible by 5.

Divisibility by 6: If the number is even and the sum of the digits is divisible by 3, then the number is also divisible by 6 (because 6 is 2 times 3). For instance, the digit sum of 54 is 9 (5 + 4 = 9), which is divisible by 3, and 54 is an even number, so 54 is divisible by 6.

Divisibility by 7: You will have to read my book *Speed Mathematics* for a full explanation of this, but here is a quick introduction. You multiply the final digit of the number by 5 and add the answer to the number preceding it. If the answer is divisible by 7, then the number is divisible by 7. For example, let's take 343. The final digit is 3. We multiply 3 by 5 to get 15. We add 15 to the number in front of the 3, which is 34, to get 49. Then, 49 is evenly divisible by 7, so 343 is as well.

Divisibility by 8: If the final three digits are evenly divisible by 8 then the number is evenly divisible by 8 (because 1,000 = 8 × 125).

Here we have an easy check similar to the check for 4. If the hundreds digit is even it can be ignored. If it is odd, add

4 to the final two digits of the number. Then, if the final two digits are divisible by 8, the number is divisible by 8.

For example, is the number 57,328 divisible by 8? The hundreds digit, 3, is odd so we add 4 to the final two digits, 28. So, 28 + 4 = 32. Thirty-two is divisible by 8, so 57,328 is evenly divisible by 8.

Divisibility by 9: If the digit sum adds to 9 or is evenly divisible by 9, the number is divisible by 9.

Divisibility by 10: If the number ends with a 0, the number is divisible by 10.

Divisibility by 11: If the difference between the sum of the evenly placed digits and the oddly placed digits is a multiple of 11, the number is evenly divisible by 11.

Divisibility by 12: If the digit sum is divisible by 3 and the last two digits are divisible by 4, the number is divisible by 12.

Divisibility by 13: Again, a full explanation is in *Speed Mathematics*. The rule for 13 is similar to the rule for 7. You multiply the final digit of the number by 4 and add the answer to the number preceding it. If the answer is divisible by 13, then the number is divisible by 13. For example, let's take 91. The final digit is 1. We multiply 1 by 4 to get 4. We add 4 to the number in front of the 1, which is 9, to get 13. And 13 is evenly divisible by 13, so 91 is as well.

Let's try it again with 299. Is 299 evenly divisible by 13? It is hard to tell without actually doing a division. We multiply the final digit, 9, by 4 to get an answer of 36. Add 36 to the 29 in front of the 9 to get 65. You may recognise that 65 is 13 × 5, but if not, carry out the operation again. We are now checking 65. Multiply 5 by 4 to get 20. Add 20 to the 6 in 65 to get 26. We can see that 26 is 2 × 13, so there is no doubt now that 299 is evenly divisible by 13.

It is often useful to know if a number has factors that can be used to simplify a fraction or to simplify a multiplication or division problem.

Appendix E:
Keeping count

People have kept count of things since the earliest times. We keep count every day. How many books do I have? How many problems have I done? How many goals? How many runs? How many points?

When I was younger I worked on electronic equipment and I had to keep count of how many units I had repaired each day. I had a goal to repair more than 80 units each day. I reached 80 several times but I never passed it.

People often keep count by drawing four vertical strokes and then crossing them out with a fifth stroke. That way it is easy to tally the amount at the end by counting in fives: 5, 10, 15, 20, 25, etc.

$$\text{卌} \quad \text{卌} \quad \text{|||} \quad = 13$$

I worked with some Japanese women at an electronics company, and they kept count by writing what they said was the Japanese or Chinese symbol for 5.

At the time I was using the wrong sequence, but Jack Zhang kindly showed me the correct sequence for this book. It is how I still keep score when I am counting various objects. The methods are useful when you have to count a selection of items. For instance, red, green, blue and yellow items that are all mixed up. I might make a list like this to keep count:

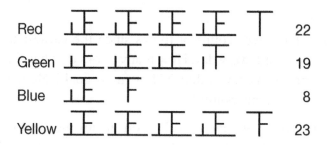

When I have finished sorting the items I simply count my symbols and write my totals to the side.

A Russian method of counting is to draw squares and then a line diagonally through when you reach five. It looks like this:

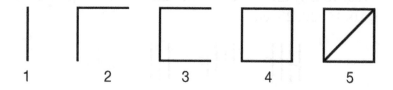

These are methods you can play and experiment with. See which method of keeping count you like best.

Appendix F:
Plus and minus numbers

Note to parents and teachers

The method of multiplication taught in this book introduces positive and negative numbers to most children. The method makes positive and negative tangible instead of an abstract idea. Positive numbers go above when you multiply; negative numbers go below. Students become used to the idea that when you multiply terms that are both the same you get a positive (plus) answer. If they are different (one above and one below) you have to subtract — you get a minus answer. Even if they don't understand it, it still makes sense.

How do you explain positive and negative numbers? Here is how I like to do it. To me it makes sense if you see 'positive' as money people owe you. That is money you have. 'Negative' is money you owe, or bills that you have to pay. Three bills of $2 is 3 times –2, giving an answer of –$6. You owe $6. Mathematically it looks like this: $3 \times -2 = -6$.

Now, what if someone took away those three bills for $2? That is minus 3 bills or amounts of minus $2. That means you have $6 more than before the bills were taken away. You could write that as: $-3 \times -2 = +6$.

I tell students not to worry about the concept too much. They don't have to understand it immediately. I tell them we will just keep using the concept and I will keep explaining it until they do understand. It is not a race to see who can understand it first. I tell the students that understanding will come.

You can give examples of forward speed and head wind. Forward speed is positive; head wind is negative. Adding and subtracting positive and negative numbers is no big deal. You add positive; you subtract negative. It is just a matter of recognising which is which.

Appendix G:
Percentages

What is a percentage?

Percentages are important. We are always meeting up with them whether we like them or not. Percentages are used in most sporting competitions. The percent sign looks like this: %.

Stores offer sales with 20% off. Taxes are quoted in percentages. We are told there is a 6% surcharge on certain items. Most people are happy to let somebody else calculate the amount for them, and they simply take the person's word for what they pay.

Often we need to calculate what we will have to pay beforehand — we need to be able to calculate percentages for ourselves.

Percentages are really fractions: 'per cent' actually means 'for every hundred'. So 50% means 50 for every 100, or 50/100. You can see that 50 and 100 are both divisible by 50. Fifty divides once into 50 and twice into 100. So

50% is the same as $^{50}/_{100}$ or $\frac{1}{2}$. It makes sense that 50 is half of 100.

If you read that 23% of people have blond hair, it means that 23 people out of every 100 have blond hair. You could also say that 0.23 of the population has blond hair, but it is easier and more common to say it as a percentage.

Money has built-in percentages. Because there are 100 cents in a dollar, 54 cents is 54% of a dollar.

Calculating a quantity as a percentage of another

How do you find what percentage one number is of another? For instance, if 32 people at a meeting are females and there are 58 people attending, what percentage are female? As a fraction they are $^{32}/_{58}$. To find the percentage you simply multiply this by 100.

You calculate this as follows:

$$^{32}/_{58} \times 100 = {}^{3,200}/_{58}\%$$
$$3,200 \div 58 = 55 {}^{10}/_{58}\%$$

This would normally be expressed as a decimal: 55.17%.

So, 25 boys in a crowd of 50 would mean that 50% are boys. Why? Because half are boys and $\frac{1}{2}$ times 100 is 50. We simply multiply the fraction by 100.

On the other hand, to find a percentage of a number — for instance, what is 20% of 2,500 — we multiply the number by the percentage divided by 100. We would set it out like this:

$$2,500 \times {}^{20}/_{100} =$$

We can cancel the one hundreds, or simply divide 2,500 by 100, which is, of course, 25. Then multiply 25 by 20 (multiply by 2, then by 10) to get 500.

$$2,5\cancel{00} \times {}^{20}\!/\!_{1\cancel{00}} = 25 \times 20 = 500 \; \textit{Answer}$$

So, summarised, to calculate the percentage of 17 to 58, for example, you would divide 17 by 58 and multiply the answer by 100. Your calculation would look like this:

$$^{17}\!/\!_{58} \times 100 = {}^{1,700}\!/\!_{58} = 29.3$$

Calculating a percentage of a given quantity

To calculate 30% of $58 you would multiply 58 by 30 divided by 100. Because 30 and 100 are both divisible by 10, you would multiply 58 by 3 and divide by 10.

$$58 \times {}^{30}\!/\!_{100} = {}^{1,740}\!/\!_{100} = 17.4 \text{ or } \$17.40$$

If you are not sure how to do the calculation, simplify the numbers to see how you do it. For instance, let's say you want to find 23% of 485 and you don't know what you have to multiply or divide; try the same calculation with easy numbers. How do you find 50% (½) of 10? You can see that you multiply 10 by ½ or you multiply by ${}^{50}\!/\!_{100}$. Then you can apply this method to finding 23% of 485. You substitute 23 for 50 and 485 for 10.

You would get:

$$485 \times {}^{23}\!/\!_{100} = 111.55 \; \textit{Answer}$$

You can check by estimation by saying that 23% is almost a quarter, and a quarter of 400 is 100.

Percentages are used in all areas of life: percentage discounts, percentage profits, statistics, sports results and examination scores. We meet them every day.

An easy short cut

There is an easy short cut for calculating percentages of round numbers.

For instance, to calculate 20% of 60 you can ignore the zeros and multiply $2 \times 6 = 12$. Twelve is the answer.

To calculate 30% of 70, multiply $3 \times 7 = 21$. That makes it easy to give an instant answer of 21.

Let's try some more.

What is 40% of 120?

$$4 \times 12 = 48$$

What is 70% of 80?

$$7 \times 8 = 56$$

What is 30% of 40?

$$3 \times 4 = 12$$

These are all easy. Try some for yourself.

How about 35% of 80?

You can simply multiply 3×8 and then add half of 8.

$$3 \times 8 = 24$$

Half of 8 is 4.

$$24 + 4 = 28 \; Answer$$

Or, you could simplify the calculation by using factors. Think of 80 as 2 × 40.

Then the calculation becomes 35 × 2 × 40.

$$35 \times 2 = 70$$

Then use the shortcut. 70% of 40 is 7 × 4 = 28, the same answer, but you are not multiplying fractions.

Or, let's try 40% of 85.

Multiply 4 × 8 and then add half of 4.

$$4 \times 8 = 32$$

Half of 4 is 2.

$$32 + 2 = 34 \; Answer$$

Keep in mind that 25% is ¼, 20% is ⅕ and 10% is ⅒.

What if you had to calculate 32% of 74? To give a close estimate, you could round off the numbers and simply multiply 3 × 8 to get an answer of 24. The actual answer is 23.68, so your estimate would be very close. This will certainly impress your friends and family and the other kids in your class.

Appendix H:
Hints for learning

We don't all think the same way and we don't all learn the same way. When I was in teachers' college, one teacher told me that if 70% of his students understood his explanation, the other students only had themselves to blame if they didn't. If most of his students understood, the others should have understood as well.

Another teacher told me, when I explain something, I can expect only about 70% of my students to understand. They don't all think and learn the same way. I have to find other ways of explaining so that the other 30% will understand as well. That has been my philosophy. I keep explaining a principle until everyone understands.

The problem is that a student who doesn't understand the teacher's explanation will generally think it is his or her own fault. They think, I must be dumb. The other kids understand, why can't I? I'm not as smart as the other kids or I don't have a mathematical brain.

The same principle applies to learning from books. A book usually has one explanation for each principle taught. If the explanation doesn't suit the way you think or make sense to you, you are inclined to think it is 'all above my head'. I am not smart enough.

You would be wrong. You need a different explanation. If you are trying to learn something from a book, try several. If your major or 'set' textbook does the job, that's great. If you can't understand something, don't think you are not smart enough; try another book with a different explanation. Find a friend who understands it and ask your friend to explain it to you. Look for other books in second-hand bookshops, ask older students for their old books, or go to your library and ask for books on the subject. Often, a library book is easier to understand because it is not written as a textbook.

When I teach mathematics and related subjects, I always read the explanation given in several books so that I can find ideas for different ways to teach it in the classroom. Also, when I am teaching a procedure in maths, physics or electronics, I do all calculations aloud, with all of my thinking out loud so everyone understands not only what I am doing, but also how I am doing it. I ask my students to do the same so we can follow what is going on inside their heads.

Appendix I:
Estimating

Often, it makes far more sense to estimate than it does to give an exact answer. Some answers can't be given with absolute accuracy; the value of pi is always approximate, as is the value of the square root of 2. Both of these values are used and calculated regularly. Even percentage discounts in your department store are rounded off to the nearest cent or nearest 5 cents. When you are buying paint or other materials from a hardware store you have to estimate. It is a good idea to estimate high to be sure you have enough nails, ribbon, or whatever it is you are buying. An exact amount is sometimes not a good idea.

We used the idea of estimation when we looked at standard long division. We rounded off the divisor to estimate each digit of the answer, then we tested each estimate.

If I am buying computer screens for a school, how much will 58 computer screens cost me if they are $399 each? To get a rough estimate, instead of multiplying 399 by 58, I would multiply 400 by 60. So, my estimation is 400 × 60, or 4 × 6 × 100 × 10, which is 24,000. Because I rounded both amounts upwards I would say I actually have to pay a bit less than $24,000. Of course, when it comes time to pay, I want to pay exactly what I owe. The actual amount is $23,142, but my instant estimation tells me what sort of price to expect.

If I am driving at 100 kilometres per hour, how long will it take me to drive 450 kilometres? Most students would say 4½ hours, but there are other factors to consider. Will I need fuel on the way? Will there be hold-ups on the freeway? Will I want to stop for a break or have a meal or snack on the way? My estimate might be 6 hours. Also, past experience will be a factor in my estimation.

The general rule for rounding off to estimate an answer is to try to round off upwards and downwards as equally as you can.

How would you round off the following numbers: 123; 409; 12,857; 948; 830?

Your answers would depend on the degree of accuracy you want. Probably I would round off the first number to 125, or even 100. Then: 400; 13,000; 950 or 1,000; and 800 or 850. If I am rounding off in the supermarket and I want to know if I have enough cash in my pocket, I would round off to the nearest 50 cents for each item. If I were buying cars for a car yard I would probably round off to the nearest hundred dollars.

How would you estimate the answer to 489 × 706? I would multiply 500 by 700. Because one number is rounded off downwards and the other upwards I would expect my answer to be fairly close.

$$700 \times 500 = 350,000$$
$$489 \times 706 = 345,234$$

The answer has an error of 1.36%. That is pretty close for an instant estimate.

Estimating answers is a good exercise as it gives you a 'feel' for the right answer. One good test for any answer in mathematics is, does it make sense? That is the major test for any mathematical problem.

Appendix J:
Squaring numbers ending in 5

When you multiply a number by itself (for example, 3 × 3, or 5 × 5, or 17 × 17) you are squaring it. Seventeen squared (or 17 × 17) is written as: 17^2. The small 2 written after the 17 tells you how many seventeens you are multiplying. If you wrote 17^3 it would mean three seventeens multiplied together, or 17 × 17 × 17.

Now, to square any number ending in 5 you simply ignore the 5 on the end and take the number written in front. So, if we square 75 (75 × 75) we ignore the 5 and take the number in front, which is 7. Add 1 to the 7 to get 8. Now multiply 7 and 8 together.

$$7 \times 8 = 56$$

That is the first part of the answer.

For the last part you just square 5.

$$5 \times 5 = 25$$

The 25 is always the last part of the answer. The answer is 5,625.

Why does it work? Try using 70 as a reference number and you will see we are dealing with something we already know.

Try the problem for yourself using 80 as a reference number. It works out the same.

So, for 135^2 (135×135) the front part of the number is 13 (in front of the 5). We add 1 to get 14. Now multiply 13 $\times$ 14 = 182 (using the short cut in Chapter 3).

We square 5, or just put 25 at the end of our answer.

$$135^2 = 18,225$$

For that we used either 130 or 140 as a reference number.

Try squaring 965 in your head.

Ninety-six is in front of the 5.

$$96 + 1 = 97$$
$$96 \times 97 = 9,312$$

Put 25 at the end for the answer.

$$965^2 = 931,225$$

That is really impressive.

Test yourself

Now try these for yourself. Don't write anything — do them all in your head.

(a) 35^2

(b) 85^2

(c) 115^2

(d) 985^2

The answers are:

(a) 1,225   (b) 7,225   (c) 13,225   (d) 970,225

For more strategies about squaring numbers (and finding square roots), you should read my book *Speed Mathematics*.

Mathematical terms (glossary)

I have written this book to try to make mathematics easy to understand. I have tried to avoid technical terms, but you still need to understand mathematical terminology. You are going to come across these terms at some point, and you need to understand what they mean.

Addend	One of two or more numbers to be added.
Common denominator	A number into which denominators of a group of fractions will evenly divide.
Constant	A number that never varies – it is always the same, like the value of pi, which is always 3.14159.
Denominator	The number which appears below the line of a fraction.
Difference	The result of a subtraction calculation.
Digit	Any figure in a number. A number is composed of digits. For instance, 34 is a two-digit number. (See place value.)

Divisor	A number which is to be divided into another number.
Exponent	A small number that is raised and written after a number to signify how many times the number is to be multiplied by itself. For instance, 3^2 signifies that 3 is to be multiplied twice (3×3). 6^4 signifies $6 \times 6 \times 6 \times 6$.
Factor	A number that can be multiplied by another number or other numbers to give a product. For instance, the factors of 6 are 2 and 3.
Minuend	A number from which another number is to be subtracted.
Multiplicand	A number which is to be multiplied by another number.
Multiplier	A number used to multiply another number.
Number	An entire numerical expression such as 349 or 12,831.
Numerator	A number that appears above the line of a fraction.
Place value	The value given to a digit because of its position in a number. 34 is a two-digit number; 3 is the tens digit and 4 is the units digit.
Product	The result of multiplying two or more numbers (in other words, it is the answer to a multiplication problem).
Quotient	The result of dividing a number by another number (in other words, it is the answer to a division problem).
Square	A number multiplied by itself. For example, the square of 7 (7^2) is 49.
Square root	A number that, when multiplied by itself, equals a given number. For instance, the square root of 16 ($\sqrt{16}$) is 4.

Subtrahend	A number which is to be subtracted from another number.
Sum	The result of an addition calculation.

Basic calculations

To picture some of these glossary terms in action, consider these basic calculations, with the role of each number alongside it.

Addition

$$23 \text{ (Addend)}$$
$$\underline{+14 \text{ (Addend)}}$$
$$=37 \text{ (Sum)}$$

Subtraction

$$654 \text{ (Minuend)}$$
$$\underline{-142 \text{ (Subtrahend)}}$$
$$=512 \text{ (Difference)}$$

Multiplication

$$123 \text{ (Multiplicand)}$$
$$\underline{\times 3 \text{ (Multiplier)}}$$
$$=369 \text{ (Product)}$$

Division

$$385 \text{ (Dividend)}$$
$$\underline{\div 11 \text{ (Divisor)}}$$
$$=35 \text{ (Quotient)}$$

Appendix L:
Practice sheets

Please photocopy these sheets as required.

Score_____

Name_____

Class_____

<u>Practice sheet one</u>

$9 \times 9 =$
◯◯

◯$6 \times 7 =$
◯◯

$9 \times 8 =$
◯◯

◯$6 \times 6 =$
◯◯

$9 \times 7 =$
◯◯

◯$5 \times 6 =$
◯◯

$9 \times 6 =$
◯◯

◯$4 \times 8 =$
◯◯

$5 \times 9 =$
◯◯

◯$4 \times 7 =$
◯◯

$4 \times 9 =$
◯◯

◯$3 \times 8 =$
◯◯

$3 \times 9 =$
◯◯

◯$5 \times 8 =$
◯◯

$8 \times 8 =$
◯◯

◯$5 \times 7 =$
◯◯

$7 \times 8 =$
◯◯

◯$5 \times 6 =$
◯◯

$6 \times 8 =$
◯◯

$7 \times 7 =$
◯◯

$3 \times 3 = 9$

$3 \times 4 = 12$

$4 \times 4 = 16$

Score_____

Name_____   Class_____

Practice sheet two

9 × 9 =   ◯ 6 × 7 =   5 × 5 =
◯◯   ◯◯

9 × 8 =   ◯ 6 × 6 =   4 × 6 =
◯◯   ◯◯

9 × 7 =   ◯ 5 × 6 =   3 × 7 =
◯◯   ◯◯

9 × 6 =   ◯ 4 × 8 =   4 × 5 =
◯◯   ◯◯

5 × 9 =   ◯ 4 × 7 =   3 × 6 =
◯◯   ◯◯

4 × 9 =   ◯ 3 × 8 =   3 × 5 =
◯◯   ◯◯

3 × 9 =   ◯ 5 × 8 =
◯◯   ◯◯

8 × 8 =   ◯ 5 × 7 =
◯◯   ◯◯

7 × 8 =   ◯ 5 × 6 =
◯◯   ◯◯

6 × 8 =
◯◯

7 × 7 =
◯◯

Score_____

Name_____

Class_____

Practice sheet three

(100) 98 × 95=   99 × 99 =   98 × 75 =

97 × 95 =   99 × 98 =   97 × 75 =

98 × 94 =   98 × 96 =   98 × 70 =

97 × 96 =   98 × 98 =   98 × 82 =

95 × 96 =   97 × 97 =   90 × 78 =

94 × 97 =   97 × 98 =   90 × 81 =

93 × 98 =   99 × 96 =   80 × 85 =

96 × 96 =   98 × 92 =   92 × 95 =

94 × 96 =   97 × 89 =   93 × 95 =

95 × 95 =   98 × 85 =   93 × 97 =

94 × 92 =   97 × 88 =   99 × 95 =

Score_____

Name_____

Class_____

<u>Practice sheet four</u>

11 × 11 =

13 × 13 =

15 × 15 =

11 × 12 =

13 × 14 =

15 × 16 =

11 × 13 =

13 × 15 =

16 × 16 =

11 × 14 =

13 × 16 =

16 × 17 =

11 × 15 =

13 × 17 =

17 × 17 =

12 × 12 =

14 × 14 =

18 × 18 =

12 × 13 =

14 × 15 =

18 × 19 =

12 × 14 =

14 × 16 =

19 × 19 =

12 × 15 =

14 × 17 =

12 × 23 =

Score_____

Name_____

Class_____

<u>Practice sheet five</u>

9 × 11 =

9 × 12 =

9 × 15 =

9 × 13 =

16 × 9 =

14 × 9 =

8 × 12 =

14 × 9 =

8 × 12 =

12 × 7 =

12 × 6 =

5 × 12 =

4 × 12 =

8 × 13 =

8 × 14 =

7 × 13 =

7 × 14 =

6 × 13 =

6 × 14 =

8 × 15 =

7 × 15 =

Score_____

Name_____   Class_____

Practice sheet six

21 × 21 =   23 × 23 =   25 × 25 =

21 × 22 =   23 × 24 =   25 × 26 =

21 × 23 =   23 × 25 =   26 × 26 =

21 × 24 =   23 × 26 =   26 × 27 =

21 × 25 =   23 × 27 =   21 × 31 =

22 × 22 =   24 × 24 =   22 × 31 =

22 × 23 =   24 × 25 =   22 × 32 =

22 × 24 =   24 × 26 =   22 × 33 =

22 × 25 =   24 × 27 =   23 × 31 =

Score_____

Name_____

Class_____

Practice sheet seven

49 × 49 =

48 × 48 =

49 × 45 =

45 × 43 =

48 × 46 =

46 × 46 =

45 × 45 =

45 × 48 =

46 × 39 =

53 × 53 =

54 × 54 =

53 × 55 =

53 × 56 =

53 × 57 =

54 × 62 =

55 × 55 =

55 × 61 =

52 × 62 =

51 × 55 =

55 × 56 =

56 × 56 =

46 × 47 =

51 × 39 =

52 × 51 =

52 × 65 =

42 × 43 =

43 × 57 =

Afterword

If you would like to learn more of my methods then I would recommend you buy my book *Speed Mathematics*. It has many more mathematical strategies not included in this book.

This book was written to explain the way I do maths. These are the methods I have developed and worked with over the years — some I was taught, some I worked out for myself. I have never looked at these methods as a series of short cuts. I often ask the question in my books and in classes, how should I subtract 9 from 56? How do I solve it? It depends on the mood I am in at the time. Sometimes I use one method; at other times I will use another. There is usually more than one way to get to the answer.

If you want to make the most of this book, use what you have learnt. Put what you have learnt into practice.

Students

People equate mathematical ability with intelligence. Use the methods you have learnt in this book and people — your parents, teachers, classmates and friends — will treat you like you are highly intelligent. You will get a reputation for being smart. That can only be good. Also, it is a fact that if people treat you as intelligent you are more likely to act more intelligently. You are likely to tackle more difficult problems.

So, use the methods. Make them part of your life, part of the way you think. Putting this book into practice can change the direction of your whole life.

Teachers

Teach these methods in the classroom and you will have high-achieving students and will yourself become a high-achieving and successful teacher. All of your students will succeed. Your class will not only perform better but they will also have a better understanding of mathematical principles. You will teach more in less time and have a highly motivated class. A highly motivated class is better behaved and will improve everyone's enjoyment of your classes. Every class will become an adventure. Everyone benefits.

Parents

Encourage your children to learn these methods. Encourage them to play with the methods. Let them show off what they can do. Never be critical of your children when they make a mistake or seem to be slow learning something. Keep telling your children they will succeed. Let them know you support them. Take their mistakes or lack of success as a challenge. Ask what you can do to make this easier. Don't compare your child with a brother or sister, or anyone else. Compare your child's performance with what he or she did in the past. Look at the improvement.

Some children in my classes come in thinking they are 'dumb'. Their self-esteem is low. These methods are an effective way to boost their self-esteem. You will find that as your children improve their performance in maths, it will have a flow-on effect. They will improve in other areas as well.

Classes and training programs

I have conducted classes and training programs around the world. I have produced material for teachers and parents to help them to teach these methods in the classroom and in the home. I often speak to home-schooling parents. You can find out more about these programs and my other learning materials by visiting my website at www. speedmathematics.com. I also have practice sheets you can download from my website.

You can email me at bhandley@speedmathematics.com if you have any questions about the contents of this book or if you would like to arrange a class.

Acknowledgements

I owe so much to so many for the existence of this book.

First, Geoff Wright, founder of Wrightbooks, who phoned me after a radio interview and said, 'Write a book and I will publish it.' He was prepared to take a risk and it paid off. Thank you, Geoff. Thank you also to Geoff's daughter, Lesley Beaumont, who gave me heaps of guidance, advice and encouragement with my earlier books.

I would like to thank the people at Wiley directly connected with the revision of this book. I would like to thank Lucy Raymond, Kerry Laundon and Ingrid Bond. They have all played an important part in the preparation of this book. There are so many others at Wiley who deserve thanks. I am grateful for your support.

Thank you to the educators around the world and to the lecturers at the State College of Victoria who have encouraged me to develop and refine my methods. I also want to thank and acknowledge the students I have taught over the years who have inspired and encouraged me to modify and improve my methods.

I would like to dedicate this book to my grandchildren and to students everywhere.

More secrets of learning from Bill Handley

Speed Maths for Kids

Speed Learning for Kids

Teach Your Children Tables, 3e

Printed and bound by CPI Group (UK) Ltd, Croydon, CR0 4YY

21/11/2022

03164642-0001